Penguin Education

Penguin Science of Behavi
General Editor: B. M. Foss

Social Psychology
Editor: Michael Argyle

The Social Behaviour of Monkeys

Thelma Rowell

Thelma Rowell

The Social Behaviour of Monkeys

Penguin Books

Penguin Books Ltd, Harmondsworth,
Middlesex, England
Penguin Books Inc, 7110 Ambassador Road,
Baltimore, Md 21207, USA
Penguin Books Australia Ltd,
Ringwood, Victoria, Australia

First published 1972

Made and printed in Great Britain by
Cox & Wyman Ltd,
London, Reading and Fakenham
Set in Monotype Times

Contents

Penguin Science of Behaviour

This book is one of an ambitious project, the Penguin Science of Behaviour, which covers a very wide range of psychological inquiry. Many of the short 'unit' texts are on central teaching topics, while others deal with present theoretical and empirical work which the Editors consider to be important new contributions to psychology. We have kept in mind both the teaching divisions of psychology and also the needs of psychologists at work. For readers working with children, for example, some of the units in the field of Developmental Psychology will deal with psychological techniques in testing children, other units will deal with work on cognitive growth. For academic psychologists, there will be units in well-established areas such as Cognitive Psychology, but also units which do not fall neatly under any one heading, or which are thought of as 'applied', but which nevertheless are highly relevant to psychology as a whole.

The project is published in short units for two main reasons. Firstly, a large range of short texts at inexpensive prices gives the teacher a flexibility in planning his course and recommending texts for it. Secondly, the pace at which important new work is published requires the project to be adaptable. Our plan allows a unit to be revised or a fresh unit to be added with maximum speed and minimal cost to the reader.

Above all, for students, the different viewpoints of many authors, sometimes overlapping, sometimes in contradiction, and the range of topics editors have selected, will reveal the complexity and diversity which exist beyond the necessarily conventional headings of an introductory course.

B.M.F.

Editorial Foreword

This volume is in the social psychology section of the Science of Behaviour series. In this part of the series a number of volumes are planned which will give a comprehensive coverage of social psychology, each written by active research workers, and providing an up-to-date and rigorous account of different parts of the subject. There has been an explosive growth of research in social psychology in recent years, and the subject has broken out of its early preoccupation with the laboratory to study social behaviour in a variety of social settings. These volumes will differ somewhat from most existing textbooks: in addition to citing laboratory experiments they will cite field studies, and deal with the details and complexities of the phenomena as they occur in the outside world. Links will be established with other disciplines such as sociology, anthropology, animal behaviour, linguistics, and other branches of psychology, where relevant. As well as being useful to students these monographs should therefore be of interest to a wide public – those concerned with the various fields dealt with.

This book takes us far beyond traditional ideas about primate social behaviour. It reports a great deal of recent research, by field studies and experiments, which is more sophisticated and detailed than earlier work. The author shows how some earlier results were artefacts of the research methods used. She is led to conclusions on topics such as territoriality and dominance hierarchies which are rather different from those currently held by many psychologists. She provides a full account of maternal, sexual and other aspects of monkey social behaviour, and relates it to biological processes. The findings on such matters as visual and auditory communication are extremely relevant to human social behaviour. The author is a leading authority on the social behaviour of monkeys, through her field studies in Africa and her highly original experimental research.

Preface

The study of primate behaviour has bred many more theories than there are facts to support them – theories which have been enthusiastically received by laymen who have no means of assessing the relative weight to be allotted to each finding. I have been working now with monkeys for thirteen years, both in the field and in cages, both as a scientist and in the practical work of making satisfactory housing, setting up compatible groups of individuals, and generally trying to keep captive monkeys healthy and contented. Thus I have become sceptical of theories unless they bear some relation to the behaviour of monkeys I know, and this book is an attempt not to present an armchair view of monkey social behaviour but to 'tell it like it is'.

To this end I have drawn preferentially, where possible, on my own work and that of my colleagues to whose animals I have been introduced and with whom I have had the opportunity to discuss their findings, rather than equally valid material with which I am not personally familiar. In this way I can have some confidence in my interpretations. I have further selected areas which I at the moment find interesting, exciting, or thought-provoking, since there is too much material available to be reviewed comprehensively in a book of this size.

Since I know the common baboon better than any other species, I have used baboons as my starting point and chief source of illustration: starting with the wild baboons I worked with, we move from this particular example to other baboons, and then to a more general account of other monkeys. For the same reason, the reader will find an emphasis on Old World monkeys, and especially the African species I know something of in their native habitat.

Although as I said there are too many theories of monkey social behaviour, I offer a few more of my own. I have not, however, referred to theories put forward in several recent popular accounts of social behaviour of primates and other animals.[1] They are readily available, and I have therefore thought it more useful to refer only to original material. Interested readers will be better able to assess these highly synthesized accounts, I hope, after reading the present one.

Species are referred to by their common names throughout, but the Latin name is also given at the first mention.

I would like to make acknowledgement to Hugh Rowell and Carol Nicoll who read the manuscript and made many helpful corrections, and to Robert Hinde who helped me especially over chapter 7.

1. Desmond Morris's *The Naked Ape* (1967) and *The Human Zoo* (1969); Konrad Lorenz's *On Aggression* (1966); and R. Ardrey's *African Genesis* (1965), *The Territorial Imperative* (1967) and *The Social Contract* (1969).

1 Perspective

What is special about primates? Human interest in animals usually has an economic basis, but this group is ecologically rather unimportant, with the gigantic exception of our own species. There is a small biomass group of nonhuman primates, and they are not very numerous except in a few forested areas, nor are the members of the group very diverse in their adaptations to their environments. And yet we are so interested in them that several species are in danger of extinction from overtrapping.

From the beginning, people have preyed on other primates. One of the earliest known sites of human activity in East Africa is a place where a group of men surrounded, killed and ate a group of baboon-like monkeys (the species is now extinct), apparently using a method still popular in West Africa today (Isaac, 1969). When people first went to Madagascar, about two thousand years ago, they found an array of unique lemuroid primates which they rapidly demolished. Only a remnant of the species remain, rather precariously, but the memory of succulent lemurs as big as calves is still alive in folk tradition there, confirmed by the bone finds of palaeontologists (Walker, 1967). This is perhaps the usual relationship between primate species of different sizes – of several species, the best documented predators are other primates; chimpanzees eat baboons, baboons eat vervet monkeys, and so on. People still eat monkeys, but the biggest users of monkeys today do not recognize them as edible. Today most of the monkeys man takes are trapped for research and drug testing, because we are primates too, and primates are more like each other than they are like other mammals. There has been time however for important differences to evolve within the group: Some of the groups of primates are separable in Eocene fossils and all the major divisions of Old World monkeys are identifiable in fossil beds about thirty million years old. On the basis of a relationship which ended thirty million years ago monkeys are used in ever-increasing numbers. Many of them could be replaced by cheaper and more easily handled laboratory animals, except that it is sometimes expedient to use 'man's closest relatives' to make research seem important. First, therefore, to bring our subject into perspective, I want to consider the basic features of primates, and to try and assess their special properties in relation to other mammals. I shall begin with anatomical features, because in the end behaviour is limited by structure, and then I will make some comparisons in behaviour.

Primates are an order of mammals, with the same status as the bats, the whales, or the rodents. Anatomically, they are rather primitive – that is they are in many ways very like the earliest mammals known. The skeletal differences which distinguish them from other mammals are practical, functional differences related to a few simple changes in way of life.

Eyes

Primates have large eyes, directed forward in their heads so that they can see things with both eyes at once and so have good depth perception. This means changes in the shape of the skull, and the proportions of the bones of the skull, to hold the eyes. Primates find their food with their eyes; probably the earliest ones hunted insects visually, as many species still do. Most mammals catch food with the mouth, and many, especially shrews and other insectivores have very strong, muscular snouts with which they can manipulate and extract the food they find. (Elephants use the same structure in a more elaborately developed form.) If you have located food with your eyes, to pick it up with the mouth you have to lose sight of it for a second, because the eyes do not focus as near as the end of the nose. You can pick it up with the hands, however, without ever taking your eyes off it – an important consideration in successful insect hunting.

Hands

The original mammals had five fingers and five toes, forming a set of very useful multipurpose tools, for holding things, digging, swimming, running, and care of fur. They can be improved for any one of these functions, but then they tend to become less good at the others. (Recall the spades of the mole, the flippers of the whale, the hooves of a horse.) As you watch an animal with specialized feet, you find that it still uses them for the other basic functions as well. Even a dog or an antelope, with feet highly modified for efficient running, will use their forefeet to manipulate objects, trying to curl the clumsy digits around and grip. Bats can actually fly with their 'hands' and at the same time catch and handle insects. Mammals with 'ordinary', relatively unspecialized hands, like rats and mice, or mongooses, all use them to hold and manipulate food and other objects.

Primate hands and feet changed relatively little, anatomically, as they specialized in holding things (*see* Figure 1). The most obvious change is an increase in the opposability of the thumb, permitting a precision grip and a power grip as well as the general wrap-around grip of unspecialized mammal forefeet. For primates, 'holding things' became a method of locomotion as they began to climb, rather than simply running about on trees. The early primates developed a method of locomotion called 'vertical clinging and leaping' (Napier and Walker, 1967) whose most advanced modern exponents, the tarsier, and the Madagascan sifaka (*Propithecus verreauxi*) are rarely seen outside their natural habitats. The sifaka (*see* Figure 2) is a white lemur weighing about two and a half kilos with a very short black face in a head it can turn right round to look over its shoulder, and extremely long hind legs. It clings to an upright tree trunk with hands and feet, its legs coiled under it like a frog's. It turns its head to fixate a trunk three or four metres away, then suddenly uncoils its legs, pushes off from the trunk, and turns in mid-air in plenty of time to clasp the next trunk. It is probably in relation to this way of moving that fingernails were evolved to replace claws, and since all primates have fingernails, it is probable that the ancestors of the whole group used vertical

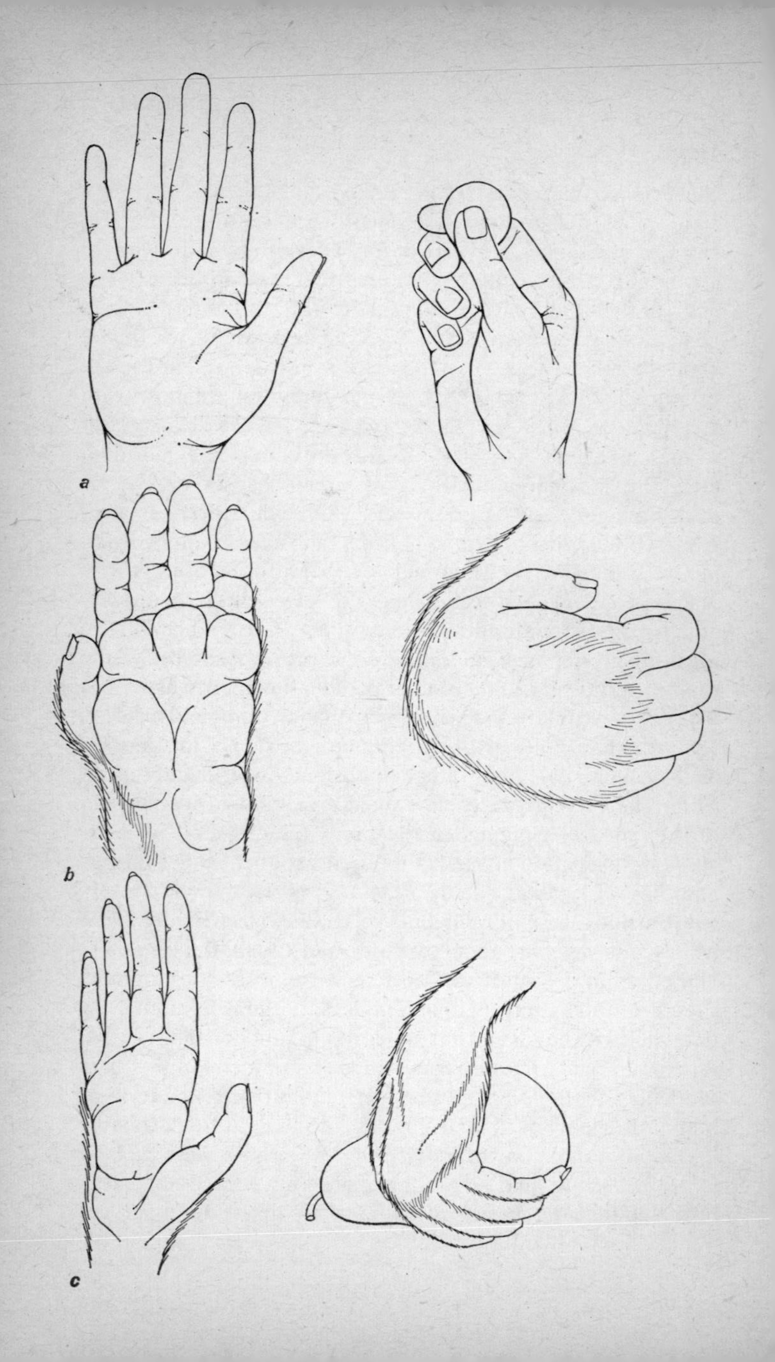
a
b
c

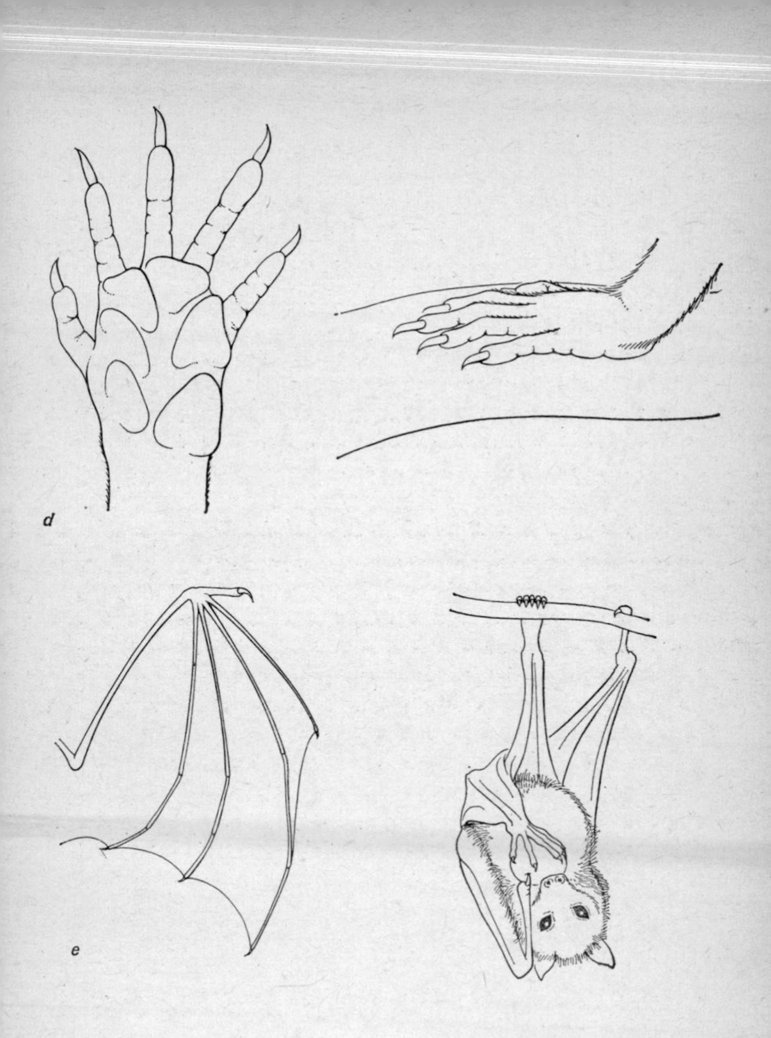

Figure 1 Primate hands have been modified for improved gripping, but are still not very different from those of earliest mammals
a Man
b Baboon
c Gibbon
d Tree shrew (a forefoot like those of early mammals)
e Fruit bat (a highly evolved hand)

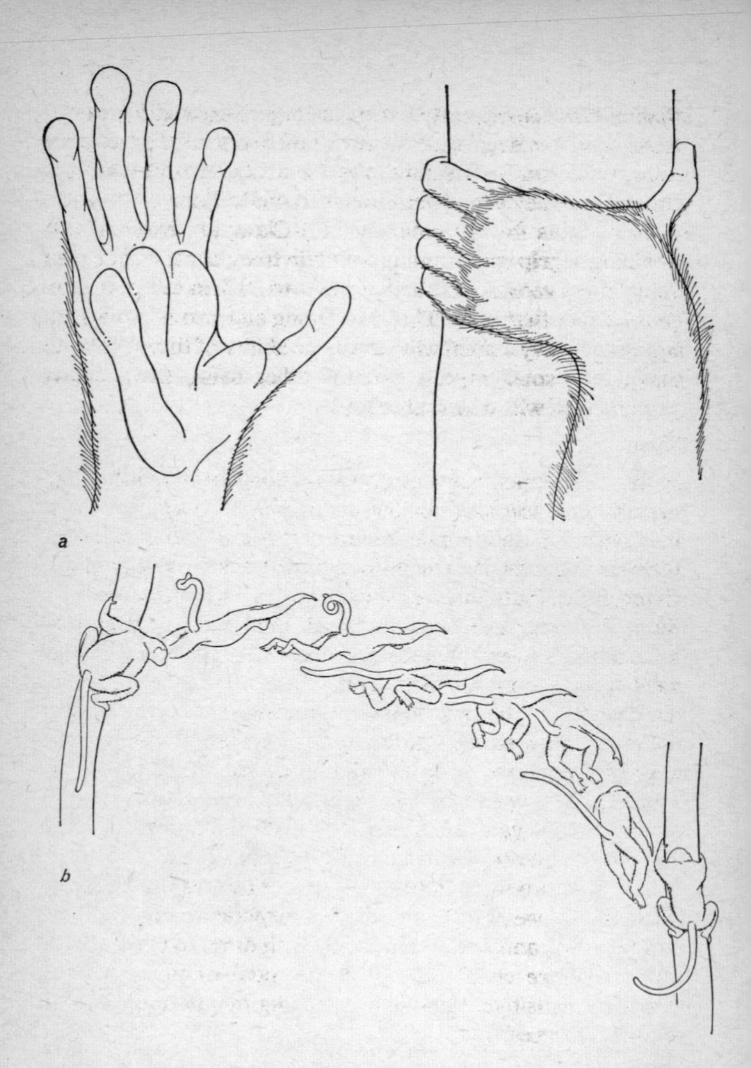

Figure 2 Vertical clinging and leaping could be the original function of finger nails. The sifaka is a lemur highly specialized in this way of moving

a Note how the hand can grip the tree

b This is drawn from a film of a sifaka's leap

Source Napier and Walker, 1967

clinging and leaping early in their history. Fingernails provide the backing for finger pads, which provide a firm grip on a tree trunk which can be instantly released at the start of a spring. (Frogs have developed very similar modifications of the ends of their digits for the same reason.) Claws are excellent for providing a grip when running about in trees, but to support an animal on a vertical surface they have to stick in to the surface (bears climb that way). Thus claws long and strong enough to support a heavy animal will reduce the ability of the forefoot to manipulate small objects. On the other hand, heavy claws combine well with a muscular snout.

Noses

As they developed hand–eye coordination for food gathering, primates had less need for the strong muscular snout of early mammals and the muzzle became short and rather delicate. (One can estimate the size of the soft snout of a fossil from the shape of the skull bones to which the muscles were attached.) Since food was not found by smell, that sense became less important – the area of sensory epithelium in the nose became smaller, and more significantly its representation in the brain also decreased: there was less of the brain devoted to correlating and associating smells. The anatomy of the nose separates the apes and monkeys from the 'half-apes' or prosimians. The latter all have wet noses like dogs and a larger nasal nervous system. But if you watch them you gain the impression that they too sample the world mainly with eyes and ears.

This is not to say that primates do not use their noses. They use them, as we know from our own experience, especially to sample food, and to find out about each other, particularly in courtship (*see* chapter 6). Even the nose of urban man is extremely sensitive, capable of detecting minute quantities of some substances.

Teeth

As the nose is reduced the face becomes shorter, and the number of teeth reduced from the original mammalian forty-four to thirty-six, and then to thirty-two in the apes and monkeys of the Old World. The first primates probably ate insects like the other early mammals, but they began to eat a higher proportion of fruit, seeds, and leaves, to chew which broader flattened back teeth were developed. Teeth are modified rather rapidly in evolution as the diet changes, and so each small group of primates has recognizably different teeth, within the basic pattern of two incisors, a canine, two or three premolars and three molars in each quarter of the jaws. Even the baboons, whose faces have become secondarily elongated, have the same thirty-two teeth, but each cheek tooth takes up a lot of jaw. The primates developed a wider range of tooth patterns than is represented by living species today – for example several of the early groups had heavy front teeth and cheek teeth rather like a squirrel.

Reproduction

Finally, primates have a unique feature in their reproduction – the menstrual cycle (*see* chapter 6). As far as is known all primates show this uterine cycle, though in fact it has been properly described in rather few species, and at the time of writing, the cycles of the New World monkeys at least are very far from being understood. The menstrual cycle apparently continues throughout the life of a female primate, except for pregnancies and a 'lactation interval' after birth of an infant. It is linked to the ovarian cycle of ovulation but not entirely dependent on it, so that outside the breeding season there may be menstrual cycles without ovulation. I do not know of any explanation of the menstrual cycle in terms of selective advantage. It may be just one of the many alternative ways of arranging an internal niche for the developing embryo which must have been 'tried out' by the mammals early in their evolutionary history, and which proved workable. (This suggestion implies a separate development of the primate line from the earliest stages of the development of placentation.) Conaway and Sorenson (1966) described menstruation in tree shrews, and speculated about the possible origin of primate menstrual cycles from the reproductive pattern of these small insectivorous mammals from South-East Asia which are in many ways intermediate between primates and insectivora.

There are other features of the reproductive process which have important bearing on the social behaviour of the group. Primates have a very long gestation for their size, and produce only one, or, in a few cases, two, young at a time, which are carried around by the mother for a period at least as long as the gestation. Some small lemurs and bushbabies provide an exception here in that they build a nest and leave their newborn infants in that, and these species may have as many as three young. The young primate matures slowly. Monkeys and apes take several years to reach sexual maturity, and then require a further period, especially noticeable in the males, to reach full social maturity, with maximum development of the secondary sexual characters.

Social behaviour

With the exception of a few of the smaller prosimians, primates live in groups. In the same way that medical science has been interested in monkeys because of the anatomical and physiological properties they share with people, some of which are discussed above, students of behaviour have been interested in them because their behaviour, and above all their social behaviour, also has special features which we and other primates share. There have been two general approaches to the social behaviour of primates, First, it has been taken as a simplified model of current human behaviour, with the advantage that monkeys have a shorter life span than we do and so can be followed through more stages during the average research project. When watching monkeys interact one is not distracted, as with people, by what they think they are doing, and how they rationalize it, but can concentrate on what actually happens. Secondly it is used as a model of stages through which the evolution of human behaviour is presumed to have passed. In this latter approach, the recent emphasis has been on trying to correlate particular social structures with attributes of the habitats in which they are observed, so from this it may be possible to deduce what environmental selection pressures are involved, and perhaps ultimately to understand the factors which led to the emergence of Man.

I have perhaps given the impression that students of primates are totally anthropocentric. Nothing could be further from the truth, and most of us work with monkeys simply because they fascinate us and we want to understand them for their own sake. But there is no doubt that the large sums of money which have been spent on research into this small and, apart from man, not remarkably successful group of mammals, of small economic importance, have been made available in the hope that it will enable us to understand ourselves better.

How different are primates from other animals in their social behaviour? Can peculiarities be related to the anatomical features described above? I should like to open this subject by summarizing two recent studies made in the National Parks of Tanzania, to provide food for thought about the uniqueness of primate social behaviour while reading the subsequent chapters on monkeys.

The first is a long-term study of zebra by Klingel (1967). He found that on the plain zebra live in small groups, consisting either of a stallion and a few mares and their foals, or only of stallions. Also there are a few solitary stallions. The groups are permanent: even when large numbers of zebra come together, as in the bi-annual migration in the Serengeti, the group remains together. The mares of a group walk from place to place in a single file, in a fixed order, led by an older mare (*see* Figure 3). The foal, and perhaps the yearling of each mare follow

Figure 3 A family group of zebra

immediately behind her, youngest first, in front of the next mare. The stallion follows the group, but can change its direction by moving up to the front and turning the lead mare with threatening gestures. In their second year colts and fillies usually leave their parents. Fillies are lured away by other herd stallions when they first come on heat – their behaviour in heat is different from, and much more obvious than, the behaviour of mature mares, and this attracts stallions from some distance. They may move through several family groups, 'stolen' again in each successive heat, but settle when they become pregnant, and are unlikely to leave the family in which their first foal is born. The mares of a family threaten a newly arrived filly at first, but slowly come to accept her. Colts wander away to find playmates, and they stay longer in their family if there happen to be other colts to make their own playgroup. They usually form bachelor groups, led by an old stallion. They are not driven out by their father; indeed he is on the best of terms with his sons, protects them, and searches for them more eagerly than their mothers should they become separated from the family. If a family stallion dies, he is immediately replaced by one from a bachelor group, or another family stallion may take the mares into his own family. Stallions don't fight over mares, however, except occasionally over the unsettled young fillies. On the contrary, they have the most cordial relationships, with special greeting and grooming ceremonies between family stallions. In one case recorded by Klingel a family stallion allowed two bachelors to join his family, and eventually went off with them to form a very stable bachelor group, leaving his mares to be taken over by a solitary stallion. Klingel used immobilizing guns to catch his animals for examination and marking, and to do simple experiments on the effect of removing one member of a social unit. He found that zebra showed great concern for immobilized friends, both mares for stallions and stallions for mares, and would wait for the 'sick' animal and lead and assist it back to the group as soon as it could rise. Also zebra that were sick from natural causes were not rejected, but continued to live with their group and derive protection from it, though only healthy mares were seen to lead a group.

Figure 4 A small pack of hunting dogs such as are seen in the Serengeti

The second study is that of Kühme (1965) on Cape hunting dogs (*Lycaon pictus*) which is perhaps especially interesting because these ungainly looking big-eared dogs have been traditionally for African game hunters the symbol of mechanical bloodthirstiness (*see* Figure 4). This study lasted for only a few months, and so leaves many questions unanswered, but once again the author recognized individuals and followed them continuously. His pack of hunting dogs, four males, and two females each with a litter of puppies, was based on the area surrounding the two holes in which the litters lived. After a special pre-hunting ceremony, the adults hunted together, except for one, and it might be any member of the pack, which stayed behind and guarded both litters. Each dog would chase a gazelle, at the same time watching the other members of the pack, and if one seemed to be gaining on his prey the rest would abandon their own and join this more successful dog – in this way they always caught the slowest, weakest member of the gazelle herd. They quickly swallowed the meat and returned home, where they were greeted by the puppies and the baby sitter begging for food, which was regurgitated for them. The same food would be swallowed and regurgitated in response to begging several times, passing round the group until the meal was evenly distributed according to need to all pack members, and everyone was satisfied. The puppies begged for suckling, using a rather different set of gestures, and were suckled equally by both bitches, once their eyes opened and they left their own dens. Begging and regurgitation seemed to be the basic behaviour pattern which held the group together, perhaps comparable, as we shall see, to mutual grooming among primates. The hunting dog pack has much in common in its organization with that of wolves (the recent book by Mech (1970) provides an excellent review of our current knowledge of the wolf, another much-maligned animal). Its organization, however, had one big difference: whereas wolves appear to be organized in rigid hierarchies of dominance, members of the hunting dog pack had no ranking, but were completely equal in status. Apart from suckling, which was of course only done by the females, all members shared in supporting the pack. Rather than their undeserved evil reputation, hunting dogs should be the model of cooperative society among mammals.

It is clear from these examples that other mammals beside primates may form groups with structure that we would recognize as having social organization, that is, they are based on individual recognition, long term association, and a high degree of mutual cooperation.

Such a level of organization is by no means universal. The minimum social life of mammals is the formation of mating associations, and of the mother and infant group. Some species, for example the hamsters, seem to limit their social behaviour to the minimum time required for these basically physiological associations. In other, socially living rodents (like rats), social behaviour appears to be determined not so much by individual relationships as by response to age and sex and similar general characters, and to the 'group smell' which identifies members of the group. The individual also has an intense, almost social relationship with the place in which it lives (*Ortstreu*). Eisenberg (1966) has reviewed the range of types of social structure so far identified among mammals.

In the same review, Eisenberg suggested some features which are generally associated with a well-developed social organization – that is something more complex than the essential mother–infant grouping, among mammals. He points out that social animals are usually diurnal rather than nocturnal, mobile, rather than sedentary, and live in grassland rather than forest. Small carnivorous members of any order are rarely social while large herbivorous species often are. These are only general indications, and there are many exceptions. It occurs to me that a common feature associated with all these specializations is an increased reliance on visual rather than olfactory investigation. As far as we know (and this may only be a reflection of ignorance about olfactory processes), the eye can more quickly perceive a more complex 'Gestalt' from a stimulus object than can nose, ears, or touch. Perhaps only the instantaneous recognition of other individuals by visual cues, and communication by visually perceived gestures, permits the development of complex, permanent societies of individuals among mammals. Good eyes, with a reliance on vision, are a basic primate characteristic, as we have seen, and if this idea is correct, primates would be preadapted to form complex social groups.

The relatively slow maturation of primates would tend to produce more complex social groups in itself. For instance, gibbons live as breeding pairs plus their immature offspring, a grouping considered a relatively low-level organization by Eisenberg. Yet a group will remain stable for five or six years on this basis, as compared with a month or so for a small rodent.

In comparison with other orders of mammals, social living is probably most common in primates, in the sense that a lower proportion of genera are relatively solitary. (The difference from the ungulates would not, I suspect, be large.) Perhaps a higher proportion of primates live in more complex types of organization, and probably the most complex forms of relationship are to be found among primates (excluding man for the moment). These differences, however, are relative, and it is difficult to pinpoint a uniquely primate feature of sociality. This judgement is based on the current state of knowledge, of course, and closer examination reveals a very heavy bias in available data. A far higher proportion of studies of primates are made on aspects of social behaviour than in other orders. We expect to find complex social behaviour in primates, and being primates ourselves we readily recognize it in them. Other animals, especially those most similar to our domestic stock, which is never allowed to develop undisturbed social structures, were expected to live in 'loose aggregations' and it has generally been stated that they do so. The zebra story given above is a good example to the contrary. Before Klingel's study, it was generally thought that zebra lived in loose unstructured herds, with perhaps short-lived male and female pairings for mating.

It is perhaps worth noting that Eisenberg's list of the qualities of social animals are also qualities which would allow a species to be easily studied by a visual animal like ourselves. Understanding social organization requires confident individual recognition of a large proportion of the population, and a long period of observation. For most species of mammals this may be impossible in the wild, especially if they are small, sedentary, nocturnal forest-dwelling carnivores. Without such a level of investigation, the complexity of social organization will be underestimated. If the observer only recognizes his animals by age–sex classes, he must assume that this is what the animal is responding to as well: he simply cannot tell whether the juvenile male is responding to the general class of aged multipare, or specifically to great-aunt Mary.

In short, the evidence that primates are unique for their generally high level of social organization is weak.

2 Ishasha Baboons

In this chapter I want to describe the life of one population of baboons that I followed for five years. This population is the baseline from which, consciously or unconsciously, I find myself assessing differences and similarities between monkeys, and so it is only fair to share this baseline before we move to more theoretical discussion. In some ways, like any real population, it was not typical, even of East African baboons, and I shall discuss its idiosyncracies and their significance later.

I went to Uganda to compare the social behaviour of wild and caged baboon groups, and to try to find a way of assessing, quantifying, and exploring the differences between them. Thus I hoped to contribute a small body of fact to the endless controversy (basically a difference of temperament) between those who maintain that behaviour can only really be studied in completely undisturbed natural surroundings and those who feel it can only be studied in electronically monitored pairs of animals in empty sound-proofed rooms.

I had chosen to study baboons because they were known to live in open grassland, where it is easy to watch whole groups and make detailed records of social behaviour over long periods of time. I was surprised to find, firstly, that in about ten years before 1962 baboons had disappeared from large areas where they had been common, under pressure of control operations and increasingly dense agriculture. Secondly, I could at first find no open grassland in Uganda of any great extent – except in areas where grass grew about one and a half metres high (the largest baboon stands not much under a metre at the shoulder). Thirdly, as I drove about the country looking at sites where helpful residents assured me they always saw baboons, I found that the places were always small cleared areas at points where the road ran through forest. I began to revise my picture of baboon ecology rather drastically, but nevertheless I settled to work in the first area in which I saw both baboons and large areas of short grass together, in the Ishasha River flats at the south end of Lake Edward, in the Queen Elizabeth National Park.

Environment

Before we move on to talk about the baboons themselves, it is necessary to have some idea of the environment in which they lived, because it is becoming clear that social organization is determined in part by the external pressures on the society from its surroundings.

In geologically recent times the Ishasha area was part of the lake bed, and the soil is a rapidly draining sandy silt. It used to be densely populated, as the immense quantities of broken pottery and the occasional human skeleton washing out of erosion channels testify. The area was evacuated in a sleeping-sickness eradication programme in the early twenties, and was incorporated only recently (1959) into the National Park. In the thirty years between, it was a game reserve, used as a hunting area by both Europeans and the local population. The short grass is grazed by famous herds of antelope and buffalo, and hippo that spend their days either deep in wallows completely covered with the floating Nile cabbage, or else in the Ishasha River.

The river is here the boundary between Uganda and the Zaïre, its other bank being the Parc National Albert. It is an unsatisfactory boundary because it continually changes its bed. It has dug a series of shallow troughs in the old lake bed which forms the main ground level, each inside the other. The present trough (*see* Figure 5) is perhaps six metres deep and a third of a kilometre wide. Within it, along the river bank, is a gallery forest, which in some places fills the river trough. This forest has a complete canopy, about thirty metres high in its mature areas, with understories of shade-loving plants. Along the edge of the river, where more light penetrates, lives another group of plants, including some very rich, dense grass well fer-

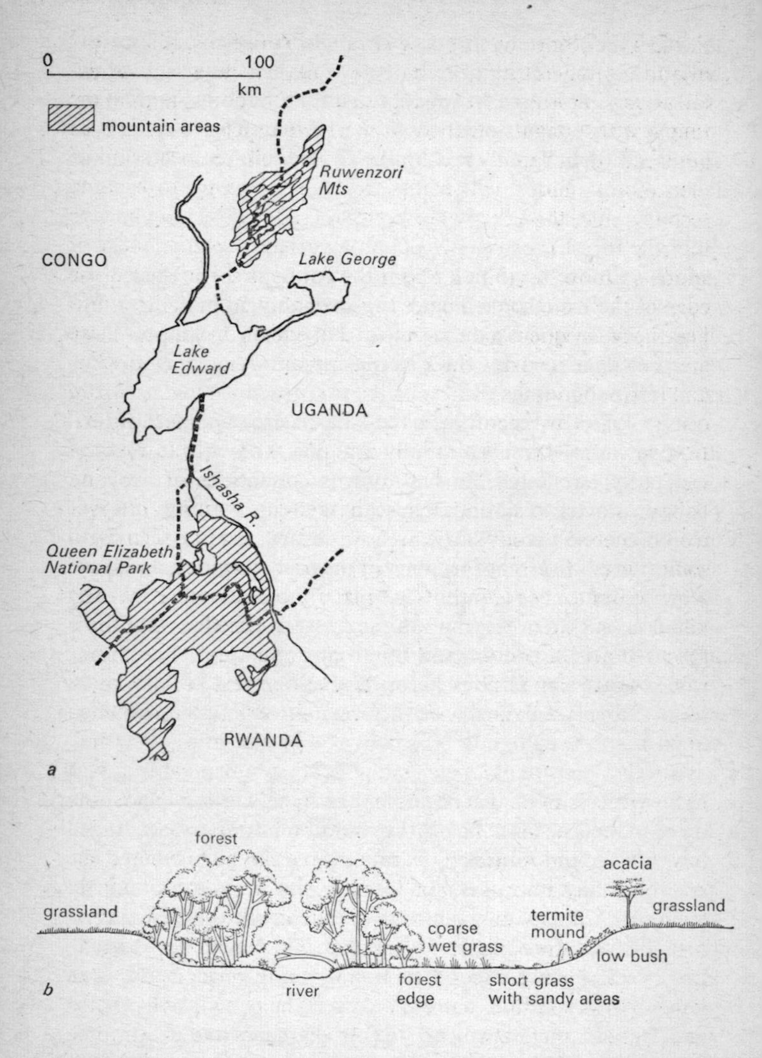

Figure 5
a **Location of the Ishasha study area**
b **Cross-section of the Ishasha River trough, showing the distribution of vegetation**

tilized with hippo manure and river silt from floods. The river meanders, undercutting old banks and undermining trees which fall across it, forming frequent bridges for baboons. At the same time it abandons its old beds which, remaining for a while as pools, are then rapidly recolonized by a complex succession of plant communities, which take perhaps ten years to become recognizable, though low, forest again. All this activity means that the forest is extremely rich in a variety of plants, most of which seem to be edible. Where the forest does not reach the edge of the trough (*see* Figure 6) – probably because of a combination of frequent burning and heavy grazing by hippo – there are patches of a strong thick siliceous grass (*Imperata*) which is edible to baboons only as it sprouts fresh after burning. *Imperata* occurs in wet areas and seems to have recently replaced forest. In drier areas there are sandy patches, with sparse tufts of vegetation, some termite mounds and occasional acacias, and baboons loved to sit and rest in these areas. Outside the river trough there is mainly short grass (up to forty-five cm high) with some herbs and scattered bushes. Aerial photographs taken a few years earlier showed that many trees had recently been killed in this area. Over a kilometre from the river there is a group of acacia and fig trees still remaining, and the baboons would make expeditions across the open grass to visit them when the figs were ripe. (I never found out how they knew when to go.)

This is a picture of an environment at a moment when it was in the process of change. Frequent burning was impoverishing the area, but the National Parks policy on burning was changed towards the end of the study, and if fires can be prevented the situation may improve. Poachers, however, will continue to burn grass where they can. On the debit side, an increasing problem was the density of elephants, which began to damage the forest. They appeared to be migrating in from the area immediately south of the Park which was being newly settled under government auspices, driving elephants and other game out of a previously wild area. The gallery forest, which was essential to the primates and also enriching the environment of most of the game, was vulnerable to all these pressures;

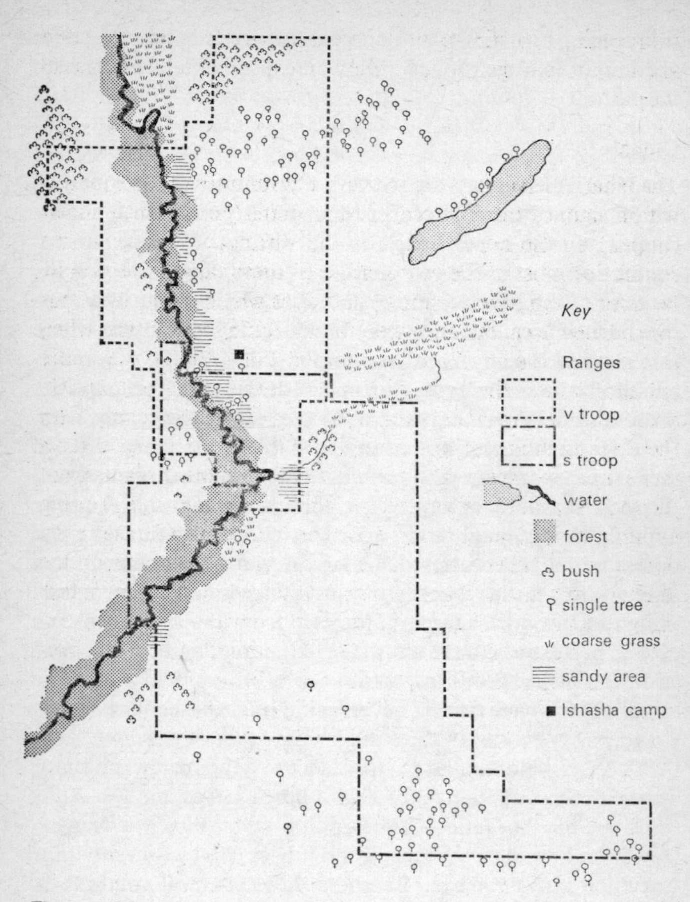

Figure 6 The Ishasha study area, showing vegetation and approximate home ranges of two troops

some of the neighbouring rivers had lost their forest fringe altogether for long stretches – it is probably difficult to re-establish such a plant community. There was also a plan, halted by the troubles in the Congo area, to build a hotel for tourists at Ishasha. If it is ever built it will probably destroy the area, unless the foraging activities of the people employed to build and staff it are far more stringently controlled than has ever been the case in the past.

Climate

The Ishasha River flats are within a few minutes of the equator, within sight of the Ruwenzori Mountains (permanently snow-capped) on the north side, and the Virunga volcanoes in the south. For most of the year neither of these ranges was visible, because of the haze of smoke and dust which hangs over this overburned area, but they were breathtakingly beautiful when rain cleared the air. Rain, as usual at this latitude, was more common about the equinoxes – in March and April, and in September to November. Some plants grew in accordance with these seasons. But on the other hand it was very rare to have more than two weeks with no rain at all, and other plants went through seasonal cycles of growth, flowering and fruiting apparently unrelated to the weather, and in some cases even unrelated to other individuals of the same species. For the animals, this meant that there was no lean time of year, something was always available. Temperature was related to cloud cover: in the wet season on a grey day midday temperatures could be below 19°C, on sunny days with little wind they might reach 36°C or so, but usually near the equator some cloud forms on sunny days around noon and keeps the temperature low. At night, the temperature never dropped below about 12°C.

Other animals

The baboons shared their home with other large mammals. In the forest were chimpanzees, and redtail and colobus monkeys. On the edge of my area there were also vervet monkeys, but they apparently did not come into the forest. On the floor of the forest, and in its edge, were giant forest hog, bushbuck, hippos, and there were also some very large pythons. In the grassland were topi, kob, and waterbuck antelopes, also buffalo, warthogs, and everywhere, elephants. There were lions, leopards, and hyenas which could have eaten the baboons, but apparently didn't. Another possible predator, of very small baboons, was the martial eagle, and vultures of four species came and ate them when they died.

Day plan

Now let us consider the lifestyle of the baboons which lived at Ishasha. Baboons slept in trees in the forest – not any tree; each troop (the word commonly used for the groups in which monkeys live) had three or four places, and nearly always slept in one of them. The troop would occupy several adjacent trees, and at least one of them at each site was unusually tall, so that its crown rose above the general level of the forest top – presumably offering a good look-out position. Sleeping places were selected with some care, older males preferring forks of large branches off the main trunk, females with infants choosing less comfortable looking, but probably safer sites among the thinner branches. Except for mothers and nursing babies, and perhaps their next-youngest child, baboons do not sleep together. Having settled for the night they usually stay in the same place till morning. Sometimes, especially on clear moonlit nights, they will perhaps spot a leopard moving through the forest, and scream and bark at it. Then a neighbouring colobus troop will join in and roar, and beneath the hullabaloo can be heard the sawing roar of the leopard. Sometimes you don't hear a leopard, and then these nocturnal disturbances sound very like the quarrels produced in the daytime when, for example, a young male teases an old female by pulling her tail. The noise made by an excited or infuriated baboon sounds to us as if it is protesting at being torn limb from limb, and I believe it is such occasions as these which have given rise to the unshakable myth that leopards prey extensively on baboons. There is no other evidence, at least for central Africa, that any of the cats take more than the very occasional young straggler.

Baboons are not early risers (which was fortunate for me because it was difficult to go into the forest before the hippo had returned to the river): though juveniles started to move about and feed soon after dawn (6:30–7:00), adult females woke later, and some adult males would still be in their sleeping places after 8 o'clock. On a damp grey rainy morning I have seen them stay 'in bed' until 9:30. Each animal on waking would scratch, then move to the nearest food – usually fruit in neighbouring trees, or in the bushes below the tree it slept in. If one of the females were sexually receptive mating also started with the first activity, and there was occasionally some roaring and chasing and manoeuvring among the males as her first consort for the day was decided on. By about half-past eight or nine the air would begin to warm up, and the baboons, having breakfasted, would move to a sunny bank of the river, or a bridging fallen tree, and adults settled to groom each other, while the infants played in nearby small trees, and older juveniles explored the possibilities of climbing a cliff or a very tall pole of a tree. After an hour or so of this one of the old males would get up and move off, striding through the grass for twenty or thirty metres and then sitting down again, facing the group. The others would glance up and notice his move, and then continue grooming. After ten or fifteen minutes he would get up again and return to the group. Another male might make a similar move in another direction. All this time the group members are not stationary – there is continual changing of grooming partners, the soliciting of still more grooming from a partner who has paused, or an attempt to get reciprocal grooming by an animal who has already groomed her partner for many minutes. There is continual jockeying for position, one animal supplanting another as groomer or groomed, either by simply pushing in between, or by approaching a smaller or less groomable animal and supplanting it. For example a juvenile will usually leave an adult female if an adult male comes and presents himself to her for grooming, but on the other hand the infant of a female in heat (oestrus) can always supplant her consort male and be groomed instead of him. Finally, one of the old females would get up, pick up her infant and, instead of moving to yet another grooming partner set off after one of the adult males who had moved out on a trail. Other females would follow her, and the group would begin to move to the next site. Usually they walked fairly briskly at this point, following traditional paths, more or less in single file.

Marching order

Once the group is under way, we should pause for a moment to consider its 'order of march'. When I was taking regular censuses of the groups, I used to identify each animal as it passed some marker bush on the trail – which was possible because they were more or less in single file. In front, of course, was the adult male who indicated the path in the first place, and after him probably one or two more big males who had suggested roughly the same direction. Then came adult females, more adult males, some large juvenile males. Then – and here counting always became most difficult – came a bunch of juveniles of all sizes, with the occasional adult of either sex. Then came more adults, with only occasional juveniles; and usually the last animal was an adult male who often seemed to be deliberately last – sitting while the juveniles filed passed him, looking carefully down the trail to make sure there were no stragglers, before moving on. The overall arrangement, then, is adults in front and behind shepherding the inexperienced juveniles between them. Small babies being carried by their mothers, of course move with adults, and so are also to be found at the front and back of a march. It is interesting that, although an adult male is usually the leader in the sense that he is at the front of the line, he is not the leader in the sense of deciding the direction in which the troop is moving. This appears to be chosen by the adult females of the group, by selecting from the suggestions of the adult males. This was possible because the baboons used only certain routes, which I also learned during my study. Some of these routes joined at some points, or had alternatives, but in some places a move in a certain direction suggested an entire day's route, with perhaps two or three possible variations. Only one prediction about the day's route invariably held true – a troop never followed the same route two days running.

The marching order I have described is actually based on an analysis of many marches, and the details of actual marches varied. The order also varied according to circumstances. One troop had a route which crossed a path used by people to fetch water from the river, and they always crossed this with great circumspection. Counts made near this path showed a much sharper separation between the sexes. Before crossing the path, the troop tended to bunch, and adult males would go quietly up the hill and peer up and down the path. If they saw people they would retire behind the hill again, and if it was all clear they would cross and the others follow. The females were more cautious even than the males, so here there would be a clear first group of males, followed by females and then juveniles, in the first part of the column.

The time at which the first march of the day started varied according to how far the group was to go that day. Sometimes they would move only about a kilometre, up or down stream along the forest edge or deep in the forest by the river and back, and then they would start later and the move might be more of a drift than a march, with some baboons foraging all the time as they walked. If they were going across the grass to the fig trees, however, they would start after only a short grooming pause, and march briskly to the trees with very little foraging on the way.

Feeding patterns

The rest of the day would be occupied with a series of feed–rest–move cycles, with the exact pattern depending on the type of food being used. A visit to a fig tree has every baboon full to hiccuping in half an hour, and is probably followed by a rather long rest and play period. On the other hand the group might move slowly along, foraging from a variety of plants as it went, with some small groups resting or grooming all the time, and then the distinction between the phases of the cycle becomes blurred. The whole group did not necessarily feed on the same thing at the same time. For example the group might rest in an open space by the river. There the females would feed on grass seed heads (plucking and pulling the heads through their back teeth to remove the seeds) while the one and two year olds climbed slender saplings and plucked young leaves from top shoots inaccessible to heavier baboons. At the same time four-to-six-year-old males would climb high into difficult trees searching under the bark for insects, and perhaps eating the nutritious cambium layer underneath. Again, later in the afternoon mothers with young babies would sit under the bushes at the edge of the forest and eat their fruit while the infants played. At the same time a foraging party of the more active members of the group, especially adult and older juvenile males and oestrous or early pregnant females, would move out into the open grass, moving fairly rapidly and apparently mainly searching for live game. They would catch grasshoppers, butterflies, lizards, and hares which would start from their forms and be coursed by several nearby baboons until caught by one. Baboons almost never take food from each other, and the captor, even if low-ranking, would tear up and eat his hare alone, unless a higher-ranking animal, by staring intently and following, could make him uneasy enough to abandon it. Such a foraging party would make a loop perhaps a kilometre long and then return to the forest edge and rejoin the rest of the troop.

Such diversification of feeding behaviour raises interesting possibilities for baboon economics. One of the theoretical disadvantages of living in large groups, offsetting the theoretical advantage of increased vigilance and security from predators, is that members of the group must compete to some extent for the available food. Baboons have such a size range that between the members of the group they can occupy a very much wider ecological niche than can any one individual, so that competition within the group for food resources is much less than one might at first think.

By the time the light is becoming less intense, about five or half past in the evening, the baboon group has probably fed on unripe tree-fruit, sweet berries, beans from acacia and other leguminous trees, growing leaves, green grass seeds, and assorted insects – a balanced diet extremely high in protein, if you compare it with a random sample of the available plant matter. At this time they are usually moving towards their chosen sleeping trees, and on most evenings they reach one of the open sandy areas within a few metres of the trees, and settle for a last bout of sociality, play and grooming, or just sitting in the sun, which lasts for half an hour or so, and then they begin to move up into the trees. Before it is too dark to see them with binoculars, most of them have settled into the places in which they will be found in the morning.

Demography

In the Ishasha area, baboon troops were strung along the river like beads on a string. They did not go more than three kilometres from their river, and so probably had no contact with the baboons of the next rivers. I knew three adjacent troops, and occasionally saw the troops up and down stream of them. Each troop used about one and a half to two kilometres of the river: no part of the area was the exclusive prerogative of one troop, though in roughly the middle of each troop's stretch of river I could be almost sure that any baboon I caught sight of would turn out to be one of that troop. I knew, however, that on the other side of the river the two end troops must meet because they occasionally exchanged members. Of the four to seven square kilometres in which I met each troop – its 'home range' – the different types of vegetation were used very unequally. (The area they used already indicated a selection of the forest habitat, because in the Ishasha area as a whole the proportion of forest or high bush is only about 12 per cent while 18 per cent of the home ranges was forest. The baboons spent 60 per cent of their time in forest or within 200 metres of it.) Because of the amount of overlap between troops it was difficult to assess the density of the population, but it was around thirty per square mile. It is also clear from this description that the troops did not have a territory – that is they did not have an area for their exclusive use or defend any area from other troops. It is possible that they had a partial spacing system: they often did some barking early in the morning at the time they began to move, and it did seem that when one troop moved upstream, its next door neighbour would also move upstream into the area they had, as it were, vacated, and I only rarely saw encounters between troops. On the other hand this idea would be very difficult to substantiate, and it was possible they all responded to generally available cues like wind, and some days were just generally recognized as 'upstream days' or 'downstream days'.

During the five years, the baboon population was increasing: there were about 105 animals in the three troops at the start and 167 at the end of the study (Table 1). Most of the increase happened in the first three years, at the alarming rate of 15 per cent per year, and the numbers remained fairly stable in the later stage. As well as increasing, the population changed its demographic structure to some extent. Since the density and structure of a population are obviously going to be important determinants of social organization, and some of the easier parameters to determine, field workers on monkey behaviour have paid considerable attention to them. Before we can understand the significance of such observations, we need to know the basic actuarial data for our species.

The female baboon ovulates once every four or five weeks. Gestation is about 180 days, six calendar months. In the Ugandan population we are considering here she does not ovulate again for five or six months after a birth – the period known as the lactation interval mentioned ealier. She typically conceives at the second or third ovulation after the lactation interval. Thus she gives birth at intervals of just over a year – about fifteen months.

The infant baboon is born with a black natal coat of very soft velvety fur, and his skin is bright pink. He looks like this for three months, and then a new coat of yellow-grey banded hair begins to come in, and his skin begins to darken. By the age of six months the coat change is complete. He continues to nurse until his next sibling is born, when he is between one and one and a half years old. Even after her new infant is born the mother protects her older child, and it helps to look after the infant. The infant weighs about 500 grammes at birth. In the fourth year, when females weigh about eleven kilos, males rather more, both sexes become sexually mature, and the females become pregnant. When the females begin to show menstrual cycles they are classed as adults by the field demographer. They do not gain their full adult dentition until about the time they have their second infant, however, when they are six or seven years old. Recent research is showing that their social behaviour has also not reached the fully mature pattern

Table 1 **Ishasha baboons**
Composition of the best-known troops in March 1965

	S troop	*V troop*
Adult males	5	14
Subadult males	0	3
Adult females:		
with swellings	1	5
pregnant	1	3
lactation interval	2	4
others	1	4
Total adult females	5	16
Large juvenile males	8	2
Large juvenile females	2	2
Large juvenile?		1
Small juvenile males	3	
Small juvenile females	2	
Small juvenile?	1	7
6 months–1 year (grey babies)	3	9
4–6 months (intermediate babies)	1	1
4 months (black babies)	2	3
Totals	32	58

at sexual maturity. Young males are quite clearly still juveniles in their fourth year, and over the next two or three years they continue to grow rapidly. (Females also grow after sexual maturity, much more if they are prevented from becoming pregnant.) By his seventh year the male has reached his full height and has his full dentition, and will be classed in the field as a young adult or a subadult male. He continues to develop in breadth and in the quality of his mane for another year or two. Fully grown, the Ugandan baboon male weighs about 40 kilos, the female about 22, though they are as variable in size as people.

We have thus an odd situation in which 'adult' is used in quite different senses for the two sexes – by the time a male is classed as adult, his female contemporary has already had three infants and is probably carrying her fourth. This situation will have an effect on the population composition reported which depends on the average life expectancy of the adult. The longevity record for captive baboons is around forty-four years. At the other extreme, Bramblett (1969) estimated about eighteen years as the life span of a Kenyan population of baboons, from an examination of patterns of tooth wear. On the latter estimate, females would be mature for half as long again as males (fifteen and ten years). If individuals lived to greater ages, the difference in age at which they are first classified as adults produces a proportionately smaller effect.

Tables 1 and 2 show the age structure of two Ishasha troops at the start of the study, and some changes in numbers of the population as time went on. In the population as a whole, the number of adult males changes less than the number of adult females, so that the 'socionomic ratio' (adult males: females) changed from unity to 0·7. The proportion of adult females to juveniles gives an indication of the rate at which the population

Table 2 **Numbers of animals in each troop through the study, and overall proportion of age–sex classes at start and finish of the 5 year period.**

	1963 August	*1964 April*	*1965 April*	*1966 May*	*1967 Jan.*	*1967 Nov.*	*1968 Oct.*
Troop							
V	45	47	58	62	57	58	64
S	30	29	31	33	30	32	29
F	? 30	45	65	60–70	73	73	74
Total	105	121	154	160	160	163	167

Start in 1963	*End* 1968
20 males	32 males
19 females	46 females
17 babies	29 babies
49 juveniles	60 juveniles
Ratios: Adult male: female .. 1:1	Adult male: female .. 1:1·5
Adult female: juvenile .. 1:3	Adult female: juvenile .. 1:2

is growing, and, as time went on, there came to be fewer juveniles to adult females, indicating that the population was growing less fast at the end than at the start of the period.

The three troops I counted did not behave in the same way. The best-known troop, *S*, hardly changed in numbers; *V* troop increased rapidly, then contracted suddenly, then grew again; and *F* troop grew asymptotically – rapidly at first and then more slowly, but steadily.

There was no difference in reproductive rate – females in all troops became pregnant at close to the maximum possible rate, and at all times of year – there was no breeding season. The differences in the three troops were produced by migration of individuals between troops.

In theory, the females in the population could have produced just over a hundred infants in five years, while the population actually increased by about sixty. Thus we have forty or fifty deaths to account for. Probably more than half of these were 'perinatal' deaths – stillbirths or deaths of very young infants – and this mortality became much higher in the second part of the study when the population growth slowed. The remaining deaths, with the exception of one young male who had had an ear abscess, were of old adults. The interesting thing is that there was no evidence at all of any baboons being killed by big cats or hyenas – predation pressure was apparently nil. (Some of the old animals could have been finished off by predators, but were clearly on the point of death when last seen.) The implication is, perhaps, that the population's numbers were being controlled by a disease affecting newborn infants, which only became a significant factor after an initial phase of rapid growth. Possibly this was a density-dependent factor which came into play as the numbers rose above some critical threshold. One might speculate that the increase in numbers was started when the Ishasha area was added to the National Park three years before the study began, thus making the area in some way capable of supporting more baboons; but it is also possible that the population undergoes long-term fluctuations in numbers such as are better known in short-lived rodents.

Movements of individuals between troops were frequent, but

not random. I knew very few individual juveniles, but the number and relative sizes of each playgroup in each troop remained about the same, and it was clear that each cohort of infants was growing through to adolescence together. In the fourth year, the picture became very different. The number of juvenile males in a troop was different at every count, and some of them must have changed troops almost every time two troops came near each other. The females of *S* troop happened to have had all male babies for two years, so at the beginning there were eight 3- to 5-year-old males and no young females in that troop. All these young males left the troop, except the one that died. There was also a net loss of young males from *V* troop. At least some of these males moved into *F* troop. I was also surprised to find that some adolescent females moved, apparently going with their male contemporaries rather than staying with their mothers; but most of them stayed in the troop in which they were born.

There was no evidence of any females moving after their first pregnancy was established. Adult males, however, moved quite frequently between troops; one of my best-known males stayed for some time in each of the three troops I watched, and by the end of the study there was no adult male that had stayed the entire time in the same troop. Some of them disappeared and then reappeared after varied intervals. One of the problems of this attempt at demography was that I was dealing with a quite arbitrary segment of the population, and it was obvious that exchanges were occurring between the troops I watched and their neighbours whom I did not know – a great deal of deduction and some inspired guesswork were therefore needed to complete the picture!

All this movement between troops occurred without any disruption of the normal pattern of life. On a couple of occasions I watched an adult male in one troop on one day and in another on the next day. In both troops each male performed the normal role of the adult male without any excitement, and if I had not been able to recognize him by the scar on his nose or the shape of his tail I would never have noticed that a change had occurred.

At this point one may legitimately ask what is the justification for identifying the troop as an entity, particularly since, on occasion, *S* and *F* troops would join forces to make the long trek to the distant fig trees.

The basis of the troops was the adult females, and home range which they worked. Their children stayed with them until adolescence, and their daughters usually for life. They were accompanied by adult males who remained with them for a few weeks or a few years. Interestingly, although the individual males changed, the number attached to each troop changed relatively little, as if there were a 'correct' number of males for each home range. When in *V* troop a lot of young males reached adulthood at about the same time and there were suddenly seventeen adult males instead of the usual eleven or twelve, seven males left the troop after a short time and the usual number of eleven was restored. If the males had been moving between troops at random, or had moved in response to particular situations (large numbers of oestrous females or very young babies have been suggested as both attractants and repellants) the numbers would not have stayed so very constant.

Roles of age–sex classes

With such frequent moves between troops, the males clearly could not establish amongst themselves any sort of rigid rank ordering or hierarchy, and indeed analysis of the few aggressive encounters between adult males did not reveal any consistent pattern of attack or avoidance between individuals. There was some tension between males, expressed by an absence of grooming between them and a high frequency of exchange of gestures of 'politeness' or 'conciliation' (*see* chapter 5). The dominant impression of interaction between males, however, was that of active cooperation. The big males policed the group. They would sit in comfortable vantage points overlooking the troop and part of the environment. If a male saw anything unusual, he would investigate or avoid, and his action would be immediately observed by other watching males, who would come to reinforce his investigation, or shepherd the troop away as appropriate. If the situation merely required 'active watching' the male would periodically make the well-known two-syllable baboon bark, often described as an alarm, but in fact mainly an advertisement of position. For example, if I came suddenly on a troop, the first response might be a true alarm bark, a shrill monosyllable from any member of the troop. A big male would then come and take up a position from which he could watch me and begin a series of two-syllable barks, spacing them further and further apart and beginning to scratch himself as he got bored and I didn't move. After a while a second male would begin to bark, becoming more distant. Baboons slipped away in the direction of the second barker, and the first male occasionally answered with another two-syllable bark. Finally only the watching male would be left. He would sit quietly for a while, and then he too would move off silently into the forest towards the distant barks. Thus much of an adult male's attention is focused on other males and on the environment. It is small wonder that the captive adult male, limited to sexual behaviour and grooming with females becomes so atrophied with boredom that he is described as stupid and sullen. Baboon males are often described as defending their troop, but this I never saw and find difficult to imagine, since Ishasha baboons always reacted to any potential danger by flight. If the threat is very small adult males may walk slowly away while juveniles run, so that they do walk between the danger and the rest of the troop. But the whole troop flees from any major threat, the males with their longer legs at the front, with the females carrying the heaviest infants coming last.

I have already mentioned the role of older females in acting as the nucleus of the troop and in determining its daily route. Adult females groom other animals incessantly: the other females, juveniles (especially their own offspring), and adult males. All these classes repeatedly solicit grooming from them, and in this way the adult females act as a focus of the group's social activity. This role is accentuated by the interest all baboons have in the young infants, that are of course with their mothers.

The adults perform the main 'service roles' in the troop. Juveniles of two years and older perhaps perform a useful function by their exploratory activity – they poke about in odd corners not visited by the adults, and find perhaps a monitor lizard, or in one case some food that I left as bait. The attention of the adults is attracted by their interest, and the adults assess the significance of a find. They would be left to explore a lizard, but when they began to taste the bait an adult male came and led them away, and it was left untouched thereafter. Older juveniles could not, however, be described in any useful sense as 'peripheral'. They were extremely energetic, so that when the main group rested they would often continue to play on some suitable nearby landmark. But the males were never rejected from the centre of the troop as suggested in accounts of other species.

3 Variations in Social Structure

The baboon lives throughout the continent of Africa south of the Sahara. Within that vast range several local races are recognizably distinct in appearance and have at various times been given specific, subspecific, or even generic rank. At present only the pink-skinned hamadryas baboon of Ethiopia is generally recognized as a species separate from other baboons, and even that interbreeds with the olive baboon where the two populations meet. (Baboons other than the hamadryas (*Papio hamadryas*) I shall refer to as *Papio cynocephalus*, following Buettner-Janusch (1966). This includes the olive forms (Ethiopia, Uganda, north Kenya, west Tanzania), the yellow (east coastal plain, Zambia) and the chacma of southern Rhodesia and South Africa.) Since baboons are relatively conspicuous and not very shy, they have been studied intensively in several places. Hall (1962a, b, 1963) studied them especially at the Cape, but also at several other places in southern Africa and in Murchison Falls Park, Uganda. Washburn and DeVore (1961) studied them in the Athi Plains near Nairobi, and in Amboseli Reserve, in Kenya. Later Altmann and Altmann (1970) studied them again in Amboseli, in much greater detail. Stoltz and Saayman (1970) investigated troops in northern Transvaal. Ransom (1971) has just finished a study in the Gombe Stream National Park in Tanzania; and a short study has just been made by Aldrich-Blake and co-workers (1971) in Ethiopia. On the hamadryas baboon, we have the studies of Kummer (1968a). At least three more studies are currently in progress in East Africa alone.

In all these places, the habitat of the baboons contained open grassland, bush, and forest, or at least groups of trees, in some of which the baboons slept. The hamadryas baboons and some

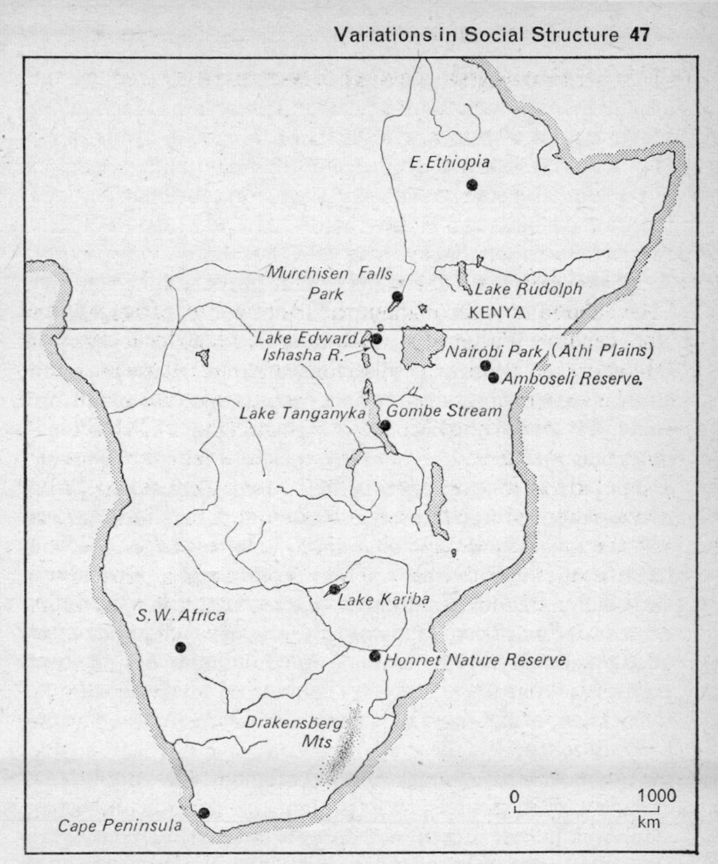

Figure 7 Location of field studies of baboons quoted in this chapter

of the South African populations slept on cliff faces. Within this general description, however, the species occupies a very wide range of habitats. Hall watched troops living on the coast, and others high in the Drakensberg Mountains in light snow. The Ishasha baboons described in the previous chapter had perhaps the richest and most forested habitat so far described, while at the other extreme the Athi Plains baboons live almost entirely in grassland. At Gombe Stream and Ishasha abundant water was always available, but in Amboseli in the dry season the movements of the troops seemed to be limited by the availability of water. Both studies in Ethiopia commented on the sparseness of food, especially in the long dry seasons, while in other areas food was regarded as abundant and varied.

These differences in habitat will have some obvious effects on the life of the baboons. The availability of food, and particularly the richness of individual sources will affect the foraging pattern. At Ishasha we saw that baboons would alternate short intensive bouts of feeding with lengthy periods of rest and social interaction. Where food is not so plentiful, and especially when it is thinly spread – for example baboons at Amboseli use the small storage bases of grass leaves as a main staple; the plants are fairly well spaced and each needs scratching up and cleaning – in such cases baboons must spend most of their day foraging, and they travel long distances in the process. Thus at Ishasha an average day's journey was two or two and a half kilometres and the longest journey about six kilometres. In the Kenyan localities the usual journey was between five and six kilometres, and again much longer distances were occasionally travelled, while in the northern Transvaal the journey was about eight kilometres a day. This meant that the amount of time and energy available for social behaviour varies inversely with the richness of the habitat. The pattern of foraging imposed by the food sources affects the spatial relations of individuals within the troop. At Ishasha much feeding could be done while seated, and a belly full was available within arm's reach. This was an alternative to the generally more usual pattern of walking slowly along plucking food between steps. In the first case individuals are much closer together than in the second. In extremely sparse habitats it becomes inefficient for the whole group to move together, and in Ethiopia Aldrich-Blake *et al.* (1971) found troops which split into small subgroups of three or four animals for much of the day while foraging. The hamadryas baboon typically forages in small groups like this in the sparse vegetation of the Ethiopian uplands.

The first response to the increasing body of information about wild primates, with the realization that a wide variation in life-styles exists among them, has been the suggestion that social organization patterns are largely determined by ecological factors such as these. Because we have several studies of baboons in different habitats to compare, we will base our consideration of this proposition on them to begin with, and later widen the discussion to include other species. The idea was first put forward as a general proposition by Crook and Gartlan (1966), and at the same time they demonstrated that an earlier assumption – that in some way complexity of social organization would be found to increase as one moved from prosimians through monkeys to apes – was not valid. Most of the data which has been used in the discussion of this proposition is not strictly about social organization at all, but about the demographic structure of populations. Troop size and composition are clearly important in determining social structure, but are by no means the whole story: it would be easy to take a given group composition and make it into several quite different types of society (*see* page 63). Since the evidence for demographic differences is rather different from, and more plentiful than that for differences in social behaviour, and demographic differences are easier to relate to ecological factors, we shall consider this aspect of social organization first.

Composition of population

To a large extent the composition of troops will depend on birth rates and mortality. We have seen in the last chapter that the general longevity of adults will affect the proportion of 'adult' males to females in the population, known as the 'socionomic' sex ratio. One of the factors influencing longevity will be the quality of food, affecting wear on teeth, and its abundance, affecting the individual's ability to recover from illness or damage. The socionomic sex ratio was very nearly one to one at Ishasha, but was nearer one to three in South Africa and the open grassland habitats studied in Kenya. Differences as wide as this would require an average age at death of no more than nine years for animals which reach maturity. Alternatively different mortality rates of males and females may be postulated. In areas where monkeys are controlled, adult males are usually shot more often than other classes if only because they present easier targets, while if any protective function of adult males could be demonstrated, this would also put them more at risk than adult females.

Juveniles are likely to be more susceptible to disease, especially if reinforced by food shortage (because of the demands made by growth processes), and also to predation, both because of inexperience and because they are a more convenient size for a wider range of predators – adults of most species of monkey are too heavy for birds of prey to lift, for example. The mortality of infants is reflected in the proportion of juveniles to adult females. At Ishasha the ratio at the start of the study, when the population was growing rapidly, was one adult female to three juveniles while at the end when numbers seemed to be more or less stabilized, it was one to two, which was about the same as that in South Africa and Kenyan grasslands.

The adult female to juvenile ratio is also affected of course by the birth rate, and this in turn is probably dependent on richness of environment, though if a population breeds seasonally the effect is less clear, because a female will either 'hurry' to join one breeding season, or leave it out altogether and wait for the next. At Ishasha females bred almost as frequently as theoretically possible, with a birth interval averaging about fifteen months. In the somewhat more seasonal climate of Gombe Stream the birth interval was about eighteen months and it was similar at the Cape, which is climatically much more seasonal, though births still occurred at any time of the year. In Kenya and southern Rhodesia birth peaks were noted, and the birth interval in Kenya was stated to be as long as two years. An annual birth for each female would give a maximum possible ratio of about one to four and a biennial birth ratio of one to two. (Because ratios of 1:2 and even higher are given for the Kenyan population I tend to doubt the estimated birth interval, since a 100 per cent survival of juveniles in a population where the adults have such a short life expectancy seems unlikely.)

These are the environmental factors which can be expected to determine the composition of a population. Overall numbers should also depend on the environment, a poor habitat supporting fewer animals than a rich one. (Note the danger of tautology here – most field workers would define the richness of the habitat in which they worked in terms of the number of animals in their population.) A smaller population must consist of either smaller or fewer troops than a more dense one, which implies either fewer possible pairs of baboons to interact within a troop, or less frequent contact with other troops, and either of these will affect the pattern of social behaviour observed. In South Africa, Hall found that baboon troops at four altitudes were progressively smaller the higher their home range – vegetation was sparser at higher altitudes. On the other hand he found that troops in Rhodesia were consistently larger than troops in South Africa, and could suggest no ecological explanation for the difference, the habitat there being apparently less rich than that at sea level at the Cape. One might make a hypothesis that where a single necessity, such as sleeping places, was limited, troops might grow and be unable to divide into smaller units because they had to continue using the only sleeping place in the area; Kummer used this idea in explaining the social organization of the hamadryas baboons he studied.

I must stress that all these ideas on the effect of ecology on the numbers and composition are nothing more than suggestions. They are based on common sense and a few observations, but even for the much studied baboon we are a long way from having evidence of causation that would be acceptable in the biological sciences. Both the samples and the differences observed are extremely small, so that the possibility that the associations which have been observed might be due to chance has by no means been excluded. To help maintain an open mind on the subject I might suggest two mechanisms by which variation in population structure might be introduced which are independent of environmental effects.

Firstly, let us suppose for the moment that there is no ecological significance to the size and composition of troops into which a population is divided, and there is therefore no selection pressure reducing the survival rate of baboons living in troops differing from the average composition. In these circumstances it would be possible for either genetic or cultural differences to arise, in populations which were geographically separated, which altered the *preferred* group size or composition but were not determined by the ecology of the habitats of the two populations. Such an explanation is perhaps less satisfying intellectually, but we have at present insufficient information to challenge it.

Secondly, although more or less equal numbers of male and female infants are born, the sex of each infant is of course independent of the others. The possible effects of this, in the composition of small groups of monkeys, was brought home to me in my own field study. One of my troops had five adult females, and each had apparently borne male infants in two successive years – statistically improbable, but not by any means impossible. The result was an entirely male cohort of three and four year olds, most of which moved to a neighbouring troop at adolescence, producing a group structure, and changes in group composition, quite unlike that which would have occurred had the expected equal numbers of males and females been born in those two years. Clearly the larger the sample of troops the less likely it is that such an unusual situation will be described as the norm, but we should not forget that the animals must have at least sufficient flexibility in their social systems to cope with such random variation in input.

Groupings

Throughout their range, and despite differences in the proportions of adult males to females, and juveniles to adult females, the olive, yellow, and chacma races of baboon live in troops of rather similar structure – each consists of several males, several females, and some juveniles and infants, all of which move around in a relatively coherent group, although temporary divisions may occur. The hamadryas baboon has a rather different group structure (Kummer, 1968a). It sleeps in very large troops on cliff faces. Looking at the animals on or near the sleeping cliff it is very striking that they are huddled in small groups, often an adult male, two or three females, and their infants. These small groups are very stable, and were described as 'one-male units' by Kummer. The name is rather misleading since in fact there is very often more than one adult male associated with the small group of females, and these males have a well-defined relationship with each other. These 'one-male units' forage together during the day, the male threatening the females if they stray and bringing them back close to him. The units are in turn associated into bands: members of a band start the day's march together, and move over roughly the same day route, but as foraging starts the component one-male units begin to separate from each other. Two or three bands share the same sleeping cliff, but may fight each other and show little friendly interaction. Kummer has traced a probable developmental sequence for the one-male unit from the variety of individual associations he saw in his study population. Briefly (*see also* chapter 7) juvenile males begin by kidnapping young infants and keeping them for perhaps half an hour. Later they adopt a juvenile female, treating her in many ways as a mother treats her infant, carrying and protecting her, but repeatedly threatening and punishing her if she strays, so that eventually she follows him closely. In early adulthood the male may acquire several females, all of which he herds energetically. The adult male often has a subadult or young adult male associated with his unit, with whom he interacts in friendly fashion. As the male grows older he spends less time herding his females, who may forage further and further away from him, and his younger male companion may also mate with his females. The older males increase their interaction with other males of the band and take a larger and larger part in determining the movements of the band as a whole while their interest in their own unit diminishes. Males have strong personal ties with other males in their band, especially with their peers, but they do not form groups consisting only of males, except for occasional pairs.

There has been a suggestion that this social structure has evolved from the multi-male group typical of other baboons in response to the difficulties of the habitat of hamadryas in Ethiopia, which combines very sparsely distributed food with infrequent suitable sleeping cliffs (Kummer, 1968b). The equivalent of the cynocephalus baboon troop is the band, which has acquired both 'fusion' and 'fission' capacity which allows it to share sleeping sites with other bands, and during the day to break up into small efficient foraging units.

In discussing this, we must first look critically at the actual differences between hamadryas and other baboons. Some of the generally accepted differences appeared to exist because Kummer's study on hamadryas was of much higher quality than the available studies of olive, yellow, or chacma baboons. In particular, he was able to collect information about interactions between individuals of all classes, whereas the studies used for comparison considered interactions largely in terms of classes of animals. More recently Ransom (1971) has completed a study of a cynocephalus baboon troop in terms of individual relationships, and found many patterns rather similar to those considered unique to hamadryas. Young male baboons kidnapped infants, adult males formed pair-bonds with older females, and developed specially intense relationships with individual juveniles. The troop could be subdivided into a series of subgroups on the basis of frequency of friendly interactions between individuals, and was altogether very far from being the rather amorphous horde of animals suggested by earlier studies. Now this is not to suggest that there are no important differences between hamadryas and cynocephalus baboons, because the briefest glance at resting groups of the two types would be enough to contradict such an idea. The differences must be analysed, however, in terms of small changes in the frequency and distribution of some behaviour patterns – a point which we shall return to shortly.

The recent study of olive baboons in Ethiopia has shed some light on the ecological explanation of hamadryas social organization. Aldrich-Blake and his co-workers (1971) found that these baboons lived in a rather poor area largely of dense thorn scrub. They slept in some tall trees along the river in coherent troops, but during the day they foraged for much of the time in small subgroups which moved independently through the scrub. The study was not long enough to analyse the composition of the subgroups in terms of individuals, though by analogy with Ransom's study we may guess there were relatively consistent companionships. There was, however, nothing comparable to the one-male units of the hamadryas. As the authors point out, since it is possible to develop a system of foraging in small groups without necessarily using the one-male unit arrangement, something other than ecological pressure is therefore needed to explain the development of the one-male unit.

There is another taxonomic subgroup of baboons, the drills (*P. leucophaeus*) and mandrills (*P. sphinx*), which we have not so far considered because they are largely unknown. They are large and very beautiful baboons which live in the tall rain forests of West Africa, and are difficult to study because of their impenetrable habitat and extreme shyness (they are hunted for food and trophies). Preliminary studies by Gartlan (1970) suggest that drills may also be organized in 'one-male' units which combine to form bands, or troops. If this impression is confirmed, (and Gartlan is continuing to investigate the problem), an ecological explanation for the one-male unit becomes rather unlikely – it is difficult to imagine two less similar habitats than the forests of Cameroon and the highlands of Ethiopia. Perhaps, as an exercise, we should stand the usual formulation of the problem on its head and ask how the multi-male group of the cynocephalus baboon has evolved from the one-male unit structure which is basic to the baboon group as a whole. The benefit of such an exercise is that it draws attention to the fact that we have no actual evidence that the more usual question is a more reasonable one to ask.

If we broaden our approach now to consider the grouping patterns of monkeys and apes in general, we find that multi-male groups similar to those of cynocephalus baboons have been found in macaque species, in one guenon (the vervet *Cercopithecus aethiops*) and in some langurs (*Presbytis spp*); and in mangabeys (*Cercocebus spp*) which are extremely closely related to baboons and in the New World in capuchin monkeys (*Cebus albifrons*) and Howler (*Alouatta palliata*). There is no logical reason at the moment to suppose that the multi-male group represents a primitive, or even a most usual grouping among Old World monkeys and apes. Examples of some alternative systems follow: in most of them the adult female is associated with only one fully grown male at a time.

The family group

This consists of a breeding pair, or sometimes more than one female and a male, and the offspring of the female(s) which have not reached maturity. Gibbons (Elefson, 1968) and *Callicebus* monkeys (Mason, 1968) live in groups of this type, and so do bushbabies for at least part of the year.

'One-male groups'

These groups are slightly larger, with several adult females and perhaps more than one adult male, but observers have usually felt that one of the males clearly had higher status than others. This arrangement seems common among the colobines – the black and white colobus (*Colobus abyssinicus*), some langurs, the proboscis monkey (*Nasalis larvatus*) and among guenons – the patas monkey, *Erythrocebus patas* (*see* Hall, 1965), and probably several of the forest guenons which have been incompletely studied. Gorillas too have groups of this type.

In a population with more or less equal numbers of adults of both sexes, groups of this type have associated all-male groups, moving separately, but within a home range largely overlapping that of one or more one-male groups. Alternatively the population may include solitary adult males, or both solitary males and male groups. (It is very difficult to be sure of a solitary male in most habitats; adult males of several species may move away from their group and stay out of sight for several hours.)

The formation of the all-male group has also been described as a response to ecological pressure. It occurs in the two monkey species able to live in completely treeless grassland, the patas monkey and the gelada 'baboon', *Theropithecus gelada* (the only surviving species of the genus, which is probably closer to *Macaca* than *Papio* on the basis of skull structure, reproductive physiology, and some behavioural criteria). It also happens, however, in langur populations in well-forested areas and in gorillas and chimpanzees, in high montane and rain-forests respectively. Crook (1966) suggested that the one-male and all-male groupings represent an economy measure where food is a limiting factor, and where the males are not sufficiently formidable to be able to protect the rest of the group. Only one male is required to impregnate several females so that most infants can be produced from a given number of adults in the one-male group structure. Surplus, or reserve, adult males live separately, not competing directly for food with breeding females. In the population of geladas studied by Crook there was some indication that all-male groups foraged more in the fringes of the habitat, and because they could move further, not being encumbered with infants, they could exploit areas effectively unavailable to breeding females. Competition for the position of male in a one-male group would act as selection pressure ensuring that only the fittest males were able to breed.

To explain the occurrence of all-male groups, solitary males, and one-male breeding groups in forest populations, Crook postulates that monkeys in tropical forest may also in fact be limited for food, in spite of its apparent abundance. The lack of seasonal variation in the quantity of food allows populations to build up to the ceiling level and remain there; similar situations have been described for populations of tropical birds. A closer inspection of tropical forests shows that the food supply is highly seasonal, at least in quality, and Owen (1969) suggests that for small birds, changes in quantity of insect food available may be big enough to provide the stimulus for migration. It will be difficult to test this hypothesis by the usual methods of quantitative ecology partly because most of the food resources in tall forest areas are inaccessible to a terrestrial primate such as an ecologist, and also because the immense complexity and variation in plant communities can be described as tropical forest. My present assessment is that the plausibility of the ecological explanation rests on the coincidence of the one-male and all-male group system occurring in both open-country species. But this has already been reported in a sufficient variety of other habitats to make a simple ecological explanation related to limited food supply rather difficult to substantiate.

Flexible subgroups

The clearest description of flexible subgroups is van Lawick-Goodall's (1968) on chimpanzees. Groups of two types were commonly encountered – maternal groups and male groups. The first consisted of mothers or a mother and her youngest children, plus sometimes older members of her family; the second might include oestrous females as well as males from adolescence on. These groups frequently changed personnel, and at first they were thought not to be referable to any definable larger group, but as more individuals were recognized it became clear that the majority of animals seen came from a local population with only limited exchange with neighbouring populations. This study was made in a rather open area (Gombe Stream National Park, Tanzania) with a relatively low density of chimpanzees. A shorter study was made recently by Sugiyama (1968) and Suzuki in Budongo Forest in Uganda, where the chimpanzee population is roughly seven times as dense as at Gombe Stream. Similar flexible subgroupings were encountered there, but the area was small enough for these workers to be able to recognize three separate groups with different home ranges. Individuals from one group would form subgroups in endless combinations; however, individuals from different groups did not form subgroups together. Aldrich-Blake (1970) studied blue monkeys (*Cercopithecus mitis*) in the same forest and found a comparable pattern of grouping: blue monkeys would be encountered in small groups of variable composition but drawn from an identifiable, larger group. The size of the subgroup depended on the food source currently being used, and occasionally the whole group would feed together in a single fruiting tree. It would then act as a group with respect to neighbouring groups, either withdrawing from the fruit tree or driving an approaching group away, depending on where the tree was in relation to the home area of each. In the New World, Klein (1971) found a similar subgrouping pattern in spider monkeys (*Ateles geoffroyi* and *A. belzebuth*). No ecological explanation has been suggested for this type of social grouping. I would speculate that as we come to know more of natural groupings of primates these types of group structure which at present seem quite sharply differentiated will become less distinct, and describable in quantitative rather than qualitative terms. We shall learn more about subgroups within the apparently tightly knit multi-male groups of cynocephalus baboons, and more of the relationships between neighbouring family groups of gibbons. The line between the macaque group in which junior adult males live 'peripherally' and the population of geladas divided into harem groups and all-male groups may also be definable in quantitative terms – or perhaps these will become two examples along a continuum.

Ecological explanations of differences in social groupings require two sorts of comparisons: of the same species in different habitats, and of species sharing the same habitat. It is axiomatic in evolutionary theory that sympatric species will have different niches; this is presumably also true for primate species, although it has not been demonstrated and is by no means obvious in some West African forest communities where several species of monkey live in extremely close association (Struhsaker, 1969). The ecological determinants so far suggested, however, have been large-scale factors which might be expected to affect all species present. Struhsaker found differing group sizes and compositions in the species in the communities he studied, and Chalmers (1968a) found mangabeys and guenons (*Cercocebus albigena* and *Cercopithecus ascanius*) living in the same forest and apparently using the same food sources, but again living in groups of very different size and composition, the mangabeys in fairly large multi-male groups and the red-tailed monkeys in small groups with often a single adult male (Haddow, 1952). When we compare the same species in different habitats, we must, as we have seen, allow for possible differences in the populations due, not to ecological factors, but to cultural or even genetic differences. Species are not homogeneous, there are local differences in physical characteristics, and so there could also be small differences in genetic determinants of behaviour. Such determinants have not yet been demonstrated in primates (they are not easy subjects for breeding experiments) but they are known in domestic dogs and rodents. With enough studies on a species, it should be possible either to recognize or discount such possibilities.

Besides the baboons, a few other species, or groups of closely related species, have been studied in three or more

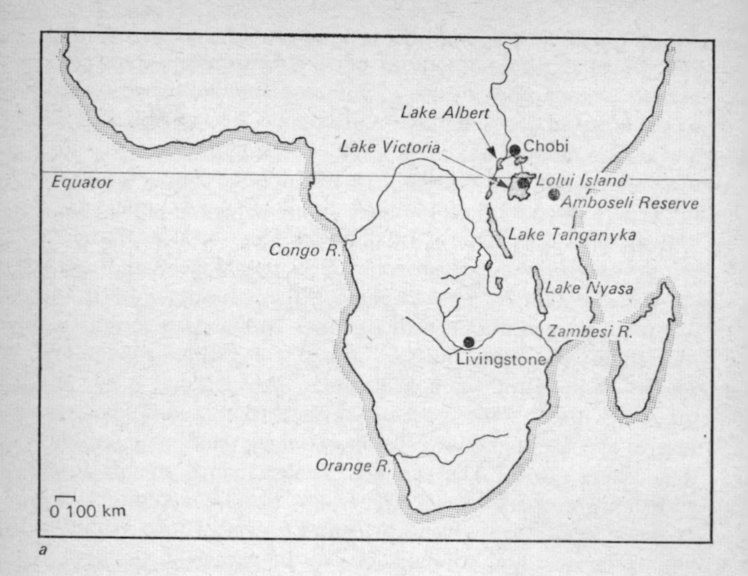

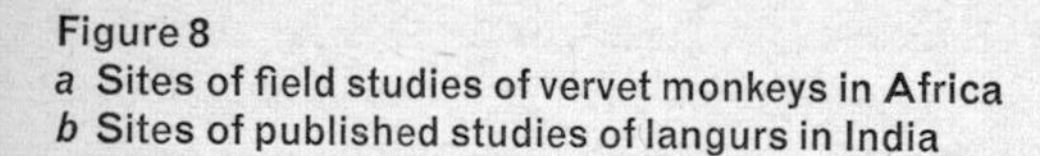

Figure 8
a Sites of field studies of vervet monkeys in Africa
b Sites of published studies of langurs in India

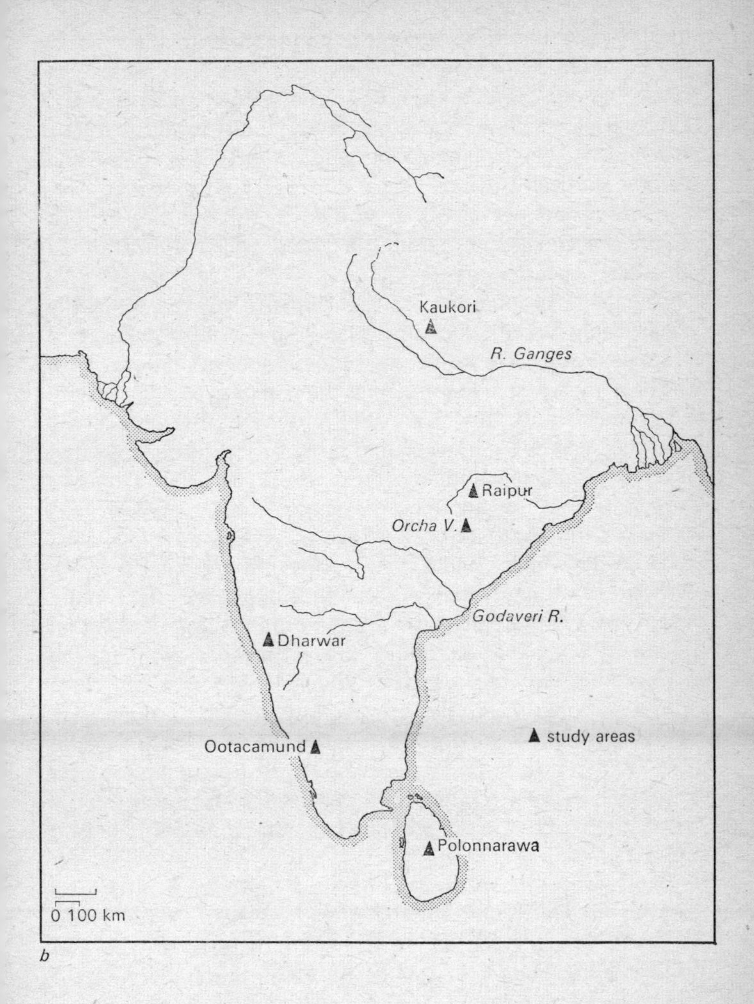
Kaukori
R. Ganges
Raipur
Orcha V.
Godaveri R.
Dharwar
study areas
Ootacamund
Polonnarawa
0 100 km

b

localities (*see* Figure 8). Foremost among these are the langurs, the common leaf-eating monkeys of India and South-East Asia. *Presbytis entellus* has been studied in five places (*see* Jay, 1965; Ripley, 1967; Sugiyama, 1964; and Yoshiba, 1968). *P. johnii*, the Nilgiri langur, was studied by Poirier (1969), and further studies of the genus are in progress. The vervet, *Cercopithecus aethiops* has been studied in three places in East Africa (*see* Gartlan and Brain, 1968; Struhsaker, 1967a) and in Zambia (Lancaster, in press).

The langur populations so far reported on were generally divided into breeding groups, plus some solitary males, and all-male groups. They differed in the number of adult males in the breeding groups: in some areas the typical group had several adult males, but a few troops had only one; at Dharwar the one-male group was typical, and most males, both adults and juveniles, lived in all-male groups. But even a few Dharwar troops had several adult males. Troop size is very variable, ranging from six to 120. At Dharwar, troops in dense forest were on the whole smaller than troops in more open country. Between localities, however, no habitat differences were found to correlate with differences in group composition (and in fact no one type of group was unique to any locality; it was only the proportions that varied). The Nilgiri langur is much more strictly arboreal than the grey langur, and its groups were generally smaller, perhaps confirming the trend in troop size seen at Dharwar, but they were also studied in an area subject to heavy human interference, where langur habitat was rapidly being destroyed and langur social structures subjected to unusual pressures.

The vervet monkey lives in heterosexual groups in all the localities in which it has been studied, and the occasional solitary male seemed to be in the process of moving from one troop to another. Troops were smaller in the more densely forested locality, and population density was also higher there. Interestingly, vervet troops are much larger in Rhodesia and Zambia than in East Africa, just like baboon troops in these areas.

To summarize the current state of knowledge of grouping patterns: there has been a wide range of patterns described, and even within species a lot of variation is possible, though there is perhaps a typical basic pattern for each. Although it seems probable that some of the variation must be related to ecological factors, in no case has a convincing relationship yet been demonstrated. There are a few parallel associations but, tantalizing as these are, it is still entirely possible that they may be mere coincidences. Traditionally at this point a reviewer proclaims that further work is required – it is indeed in progress. But I suggest that we now have enough studies to be able to say that there are unlikely to be generalizations relating habitat and grouping that will hold for all primates.

Social organization

A description of the composition of a social group by no means describes its social organization. To illustrate this we might compare the one-male unit of the hamadryas with the one-male group of the patas monkey, whose composition may be very similar. In the hamadryas group the adult male is the focus, and females interact relatively little with each other except when one of them has a new infant. The male actively herds his females, responding with threats to their straying. They in turn respond to his threats by approaching him, and if they do not, he chases and punishes them by biting into the dense fur at the nape of the neck: they then follow him back to his 'harem'. By contrast, in the patas group females interact mainly amongst themselves, and the male is rather peripheral in both the spatial and the social sense (while performing functions of the look-out and protector of the group by diversionary displays). Threats and aggressive contact are extremely rare in wild patas groups, and males do not threaten females. In a caged group threats were more common, and there females responded to male threats by the more usual patterns of avoidance (Hall, Boelkins and Goswell, 1965).

To describe the social organization of a group so that it can be compared with that of other groups, we need, beside its composition, quantitative data on the interaction patterns between members of the group. We need the answer to the questions 'who did what to whom, how often and when?'

This is in any case, the form in which information is collected. Too often, however, the material is presented in digested form. Instead of describing the interactions observed, students have written in terms of subjective concepts such as 'leadership', 'dominance', 'protection', 'relaxed' or 'intense' societies, 'friendliness' or 'aggressiveness', without these qualities ever being defined in terms of the behaviour that was observed. These subjective concepts are necessarily based on the distribution of communicative motor patterns and the response to them, as analysed by the brain of the observer. The human brain is extremely skilled at exercises in cross correlation such as this entails – even when dealing with quite complicated sets of data one usually notices, during the original observations, the majority of correlations which later prove to be significant statistically. But if the brain is not forced to state what criteria are being used in forming its impressions, objective communication about them becomes extremely difficult. If behaviour is reported in subjective terms, there is no way of refuting the charge that differences in behaviour reported in various field studies are merely reflecting differences in the bias or interests of the observer. Zuckerman (1963) suggested that the temperament and the sex of the observer might well be important filters in determining, for example, the amount of agonistic (aggressive and fearful) behaviour that was reported. Objective reporting of behaviour does not eliminate the possibility of observer-bias, of course, and in trying to record the complexities of interaction within a group of monkeys some selection is inevitable, but it does make it easier for the reader to detect such a bias.

There is perhaps a fear that social behaviour presented in terms of what is actually observed would be totally indigestible. In fact it is the presentation in terms of subjective concepts which becomes unreadable, as soon as the student has realized that a concept like 'dominance' *is* a concept and not an observable and measurable biological fact, even though it is very often accepted as such. (Some concepts which have been used in describing social organization are discussed in chapter 8.) It has been a major advance in sociology to realize that, as in other sciences, observations must be separated from deductions if either are to be of any lasting value, and that emotive words should be restricted to the latter section.

In saying that, in order to describe social organization, we need data on both group composition and the frequency and distribution of communication patterns, we are making certain predictions not all of which have yet been tested by observation. To illustrate some of these let us consider the motor patterns of threat (*see* chapter 5) in some imaginary groups of monkeys.

1. We expect that the frequency of interaction will affect social organization: thus a group in which five threats an hour are recorded will be organized differently from one in which fifty-five an hour are seen, even if the contexts in which they occur and the classes of animals involved are the same.

2. Behaviour occurring now is influenced by that which occurred some time ago: thus in a group where a high rate of threatening occurs only in certain feeding conditions, behaviour at other times will to some extent be affected, and the resultant social organization will reflect the overall level, as well the highest and lowest rates observed.

3. A group in which threats only occurred between adult males would have a different type of organization from one in which they only occurred between adult females, even if the frequency and context of threatening were the same.

4. A group in which threats between individuals were highly consistent in direction (animal *A* always threatens animal *B* and never vice versa) would be different from one in which individuals threatened each other equally often, even though the frequency and context of threatening were the same in both.

If we ask questions relevant to testing hypotheses of this sort, the majority of field studies which have been published are not helpful. To some extent, field studies are improving in the quality of behavioural data they produce. Studies which pioneer a new level of excellence, such as that of Kummer (1968a) quoted earlier, set higher standards for those which follow. There is a limit, however, to what can be achieved working with groups of monkeys whose only modification is habituation to the field worker. Even with animals so tame that they may be touched there are problems in collecting quantitatively reliable data because you cannot guarantee seeing the same individuals equally well on regularly repeated occasions. The most glaring gaps in our knowledge concern arboreal species living in dense forest where merely finding the monkeys and keeping them in sight requires enormous effort. Field-studies under these conditions are extremely uneconomical in terms of information acquired per unit of time or effort, although they are essential in that some of the information they provide can only be obtained in that way. Other types of information can perhaps be more easily obtained from situations which are to varying extents manipulated or controlled by the observer. Such studies will be the subject of the next chapter. Some of them provide indirect evidence of the environmental effects on social behaviour which direct observation in the field has so far failed to do.

4 Comparison of Field and Captivity

At the end of the last chapter I pointed out the difficulties of collecting statistically satisfactory amounts of information on social interaction in most habitats where monkeys normally live. In some places field workers have been able to bribe monkeys or apes to be found more reliably, to become less shy and so to behave more visibly, by providing a regular food supply within the animals' original home range. This ruse has been given the stylistically regrettable name 'provisionization' which we shall not use. The best-known studies using this method are those of the Japanese macaque, *Macaca fuscata*, at Takasakiyama (and now at other localities also) and that of chimpanzees at Gombe Stream National Park in Tanzania. Monkeys have been captured and released in another, more convenient area, and there provided with most of the food they need; the most famous project of this type is that on Cayo Santiago Island in Puerto Rico, where rhesus monkeys have now lived for about thirty years.

It is convenient to have terms to separate these studies from those of undisturbed primates. I shall use 'wild' to describe the latter, and also animals which take very small amounts of bait, and 'free-ranging' for animals which are not confined but which are fed almost all their food by people. (This dichotomy is not followed elsewhere in the literature – American workers commonly use 'free-ranging' for all animals not in cages – but then the need for the distinction is also not universally recognized.)

Social behaviour is also studied in groups of monkeys confined in very large enclosures – for example at Oregon Primate Research Center there is a troop of Japanese macaques which was captured in its entirety, after preliminary observation in the wild, and which now lives in a hectare enclosure. Groups have been studied in smaller runs – of the order of 5 square metres per monkey – in which it is possible to catch and handle individuals regularly, to take vaginal samples to follow reproductive cycles, for example. Many of my own observations have been made under these conditions. Other studies have been made in cages roughly an order of magnitude smaller, still using relatively long-term groups; and finally, we must also consider studies of social interaction between individuals placed together in a small space only for the duration of testing periods, and otherwise held separately. Roughly speaking, each of these successive levels of approach exchanges a further distance from naturalness in return for increases in convenience and in the amount of control the observer has over what particular information he will be able to obtain.

Clearly the choice of conditions of observation should be carefully matched to the problem in which the investigator is interested, having regard to the effect on social behaviour of each of these methods of contacting the animals. In fact there is very little information on any such effects and in this chapter we shall be forced to mix evidence with inference.

It can be argued that monkeys are typically capable of exploiting very varied habitats – they are highly adaptable. Any habitat in which the species can survive and breed sufficiently successfully to maintain itself may be regarded as part of its range, including some of the more extremely manipulated forms of captivity. At the other extreme, it is doubtful whether there are any areas of the world in which the environment is totally uninfluenced by human activity, so that differences between captive and wild conditions are only of degree. As a start, then, we will consider a situation not so far mentioned – that in which monkeys have entered into some form of commensalism with people without coercion. Isolated incidents of groups of monkeys or even apes using rubbish dumps as their main food source, or learning to take food from tourists, occur wherever human activities extend suddenly into previously more or less uninhabited areas – for instance baboons or vervet monkeys in many of the game parks of Africa, or chimpanzees at a township rubbish dump in Sierra Leone (where they had a much higher incidence of tooth decay than did chimps subsisting on a more natural diet in nearby forest). Such adaptation has been more frequent, and more intense, among macaques than any other group. The most obvious example is the rhesus monkey in northern India, and the bonnet monkey (*Macaca radiata*) in the south, but the Japanese macaque also thrives in temple gardens, and is fed by tourists, while in Gibraltar another species, *Macaca sylvana*, has achieved a similar tolerated or protected status.

A very large proportion, perhaps a majority, of rhesus macaques live in or near towns and villages, obtaining some or all of their food by theft from the more or less tolerant human population. Singh (1969) has studied the behaviour of rhesus monkeys caught in towns and compared them with monkeys caught in the forest. His results were foreshadowed by the fact that he had to pay more for town monkeys, and even at the higher price trappers were unwilling to work in the town because they said the forest monkeys were more stupid, and easier to catch. In this they were not correct, since town and forest monkeys performed equally well in various standard learning and discrimination tests, once the forest monkeys had become tame enough to work.

Singh did, however, find big differences in social behaviour. When he tested them in the field, by putting out food, he found that in the forest a single large male would monopolize the food, while the rest of the troop hovered at a distance. When tested in the breeding season, the male would permit his current consort and her most recent infant to share the food with him. Town monkeys, on the other hand, were more tolerant, and the food would be shared by the male with several females and their offspring, and their behaviour was the same in or out of the breeding season. Note, by the way, the difference between rhesus monkeys and many other species even in the forest: most species of monkey would not immediately accept bait like this, and at Ishasha, far from monopolizing proffered food, the big male baboon led investigating juveniles away from it. Again, on the Athi Plains in Kenya where attempts were being made to trap baboon in baited drop-traps, adults would sit by the cage and threaten hungry juveniles away from the bait (Maxim and Buettner-Janusch, 1963).

In the laboratory, Singh tested town and forest monkeys which had all lived in the laboratory for many months. He found that in competition for food, town monkeys always succeeded against forest monkeys. Among rhesus monkeys, adult females are usually strikingly subordinate to adult males, but in these tests female town monkeys would take food before male forest monkeys. When monkeys strange to each other were put together in small rooms, forest monkeys would settle down with relatively little fighting. Town monkeys would attack each other or a forest monkey fiercely, and some town monkeys were even killed in such tests. What are the differences between these two habitats, that may perhaps have caused these very large differences in social behaviour? The visual stimulation they provide is very different, and Singh also showed differences between the two groups of monkeys in their 'curiosity', in tests where they had the opportunity to observe scenes through windows in their cages: town monkeys were more curious to look out of the windows and more interested in complicated, moving views. The most obvious difference, however, is in the sort of food available, and the strategies required to obtain it. The forest monkey uses knowledge of the plants in its home range to find food, but once it is found there is normally plenty for every member of the group to forage at leisure. Food is scarcer in the town (town animals usually look in poorer condition) and the monkeys compete for every scrap, both among themselves and with the people and their livestock. It is not surprising that the town monkeys were better at competing for food. What is more interesting is that they were also so much more aggressive in situations which did not involve food, and that the difference remained even after they had been kept in the laboratory and fed regularly and without any necessity to compete for more than a year.

When people provide food for monkeys, either in their natural habitat or in captive groups, it is usually provided in one place, or only a few places, and relatively rarely – perhaps once or twice a day. When people feed monkeys to amuse themselves the feeding schedule tends to be even more erratic, and the quantities small. All these factors tend to introduce an element of competition into acquiring food which is rarely present in normal foraging situations. When eating grass or berries, a monkey is primarily interacting with his environment, concentrating on finding his next handful, for himself. Of course, social behaviour is not absent – he is eating with other animals, his choice is probably influenced by social facilitation, and he has learned to recognize the food in the first place by observing his mother or siblings (all these processes have been demonstrated in laboratory tests). When taking food from people, social interaction becomes much more prominent. Each manoeuvres to place himself in a better position relative to the food source, and can increase his chance of obtaining food by attacking others and driving them away. In some situations additional stress derives from the element of fear in approaching a person or apparatus from which food may come.

Chalmers (1968b) observed some 'natural experiments' of the effect of food source on social behaviour when watching white-cheeked mangabeys, in which the complicating factor of human intervention was eliminated. On some occasions the mangabeys were feeding in trees with small, scattered fruit – rather like a huge cherry tree; on others he watched them feeding on jackfruit, enormous warty fruits which grow directly on the main stem of the tree, and only one of which ripens on the tree at a time. Chalmers recorded several kinds of aggressive interaction in these two situations, and found that aggressive behaviour of all kinds was about eleven times more frequent when the animals were working on the localized food source. Most of this was not a matter of direct competition for pieces of food, but seemed as he suggested to be generated partly by the unusually close proximity into which the monkeys were forced. This finding could be corroborated, though much less elegantly, from the Ishasha records. Two types of limited food were seen there, mushrooms and seedlings which grew out of mounds of elephant dung. Aggressive behaviour among the baboons was far more frequent and more intense over these productive piles of dung than in any other situation.

There are two reports of the effect of severe food shortage on social behaviour, that of Hall (1963) on chacma baboons, and that of Loy (1970a) on the Cayo Santiago rhesus monkeys, and their observations are very similar. The chacma baboons had been marooned on an island in the rising waters of the Kariba Dam, while at Cayo Santiago the regular chow ration did not arrive. In both cases there was a sharp drop in all forms of social behaviour, both friendly and agonistic. The animals foraged continually, scattering and breaking up into small groups to do so. Sexual behaviour decreased among the rhesus (who were only hungry for three weeks) and was completely absent in the baboons, where no females showed any signs of the swellings which accompany oestrus in this species. The baboons did not play, and the rhesus hardly did so.

It seems reasonable to conclude that food supply, both the quantity and the form in which it is available, has a direct effect on the amount and type of social interaction seen in monkey groups. This has not generally been taken into consideration when using data from artificially fed groups and comparing them with undisturbed animals, and this has led to some probably unjustified conclusions in the past. For example, it is much easier to feed monkeys which normally forage partly on the ground, as compared with those which forage in the treetops. Thus our only data on arboreal species derives from undisturbed groups, while at one time practically all the available information on more ground-living species derived from groups which were artificially fed in one way or another. It seems quite possible that many of the characteristics which had been thought to be associated with coming out of the trees and into open country (DeVore, 1964) might in fact be related to artificial feeding. One such character is a high degree of aggressiveness and an obvious hierarchy among adult males, which were described for macaques and for baboons which were fed, but not seen in the Ishasha baboon troops which were not. Another pattern of social organization which might be queried on these grounds is the concentric arrangement of animals in a group. This again has been described in fed baboons (Washburn, DeVore, 1961), and Japanese macaques (for example Sugiyama,

1960), and in urban rhesus monkeys. There is a central, alpha male, surrounded by females with young infants, the highest ranking in the centre. Juveniles, especially males, are more peripheral, and there are also peripheral, low-ranking adult males. This pattern was said to be clearest in the baboon group when the animals moved; yet it is very different from the marching order seen at Ishasha and described in chapter 2. It is perhaps significant that Singh used the concentric picture to describe the results of his feeding tests on wild monkeys. Lindburg (in press) observed both forest rhesus monkeys and groups which foraged in a nearby Research Institute complex, and was unable to detect the concentric pattern in the forest groups. I suggest that the concentric arrangement of monkey troops may be an effect of artificial feeding, and that the circles are not centred on the alpha male so much as on the desirable food source he has monopolized. This hypothesis will be disproved when a concentric arrangement is described in a non-fed monkey group.

When we turn to caged groups of monkeys, there are additional factors to consider. Caged monkeys are forced to be closer to each other, *on average*, than they would be in the wild, where though they would probably spend a lot of time in contact with other monkeys, they would also on occasion be many metres apart. A simple description of the cage in terms of available square metres is probably not enough because one finds in practice that a cage subdivided into smaller areas is more satisfactory than a single large space. If there is a mesh partition between two animals they are effectively further apart than if they sat at the same distance without the mesh, even though they can see, hear, and smell each other equally well. This is why animals in zoos come to tolerate people and even feed from them, although very shy. Some species are not able to accept bars as 'effective distance', and remain shy; they are not popular exhibits. Hediger (1955) discussed these problems of space in terms of 'flight distance' and 'individual distance'. The former is the space an animal maintains between itself and a potential enemy. If the dimensions of a cage are not larger than the flight distance, the animal repeatedly hurls itself against the opposite wall of its cage, and is unlikely to survive. Each species has a typical flight distance, but it may be modified by learning – for instance learning the value of barriers – and this is the process we should call taming. 'Individual distance' is the space an animal normally maintains between itself and other members of the species: it is again species specific, and not so easily modifiable by experience. There are big differences in individual distance between even closely related monkey species: Rosenblum, Kaufman and Stynes (1964) kept groups of pigtail and bonnet macaques (*M. nemestrina* and *M. radiata*) in identical cages, and showed that the amount of contact each individual had with others in the group was quite different in the two cages. Bonnet macaques normally huddle together in clumps, pigtails sit separately. Even when they groom, pigtails make only the one hand-to-fur contact. Even species which normally maintain a large individual distance must make some contact with each other, if only for mating, and they usually have rather elaborate series of gestures which accompany approaches and serve to reduce the tension the situation arouses. Species which contact frequently and extensively do so with little preliminary. Primates as a group, however, have short individual distances when compared to other groups of mammals.

Most monkey cages are rather bare and open. This is practical because hygiene is an important factor in monkey maintenance, and the more complicated the shape of a place the more difficult it is to clean. This usually means that animals in a group can see each other all the time, which is unlike conditions in even the bleakest natural habitat. Caged groups are occasionally completely disrupted, and animals die, because they seem to become trapped in a spiral of fighting. This doesn't happen in the wild, as far as we know, because a threatened animal simply moves out of sight behind a bush until the incident is forgotten. There is a hint that even in the wild the amount of cover may affect the amount of agonistic behaviour seen. Studies of baboons in open country have stressed aggressive behaviour (Altmann and Altmann, 1970; DeVore, 1965), while in baboon studies in denser vegetation the small amount of aggression seen has been specifically commented on (Hall, 1962b; Rowell,

1966a). For caged groups, cover can be provided without sacrificing cleanliness or visibility by a series of solid partitions so arranged that the observer can see into all segments, but an animal in one segment cannot see those in another – a cage in the form of a segmented arc with the observer stationed at its centre (*see* Figure 9). With gates through the partitions controlled from the observation point, such an arrangement also makes manipulating animals easy.

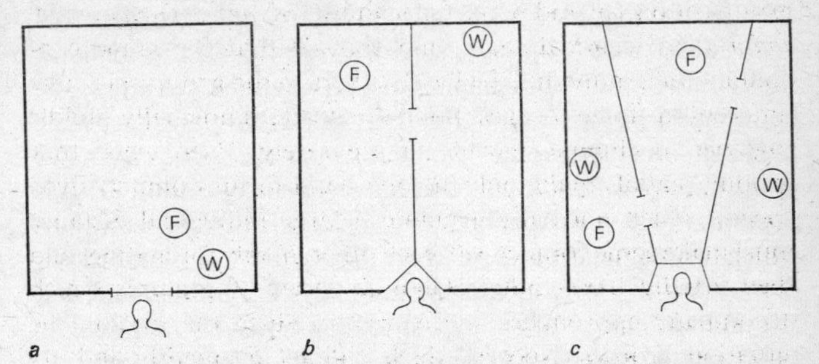

Figure 9 The use of space to reduce stress for caged groups of monkeys
a **Undivided area. Single source of food (F) and water (W) can only be reached by approaching the observer**
b **The same area divided into two by a solid partition, so that members of the group can be out of each other's sight. Food and water away from the observer. This arrangement can still be dominated by an animal in the doorway between the sections**
c **The two partitions here make surveillance by one animal impossible. The group can subdivide in more complex ways. More than one source of food and water reduces competition for these commodities**

Another difference between caged and wild groups is the initial relationship of the animals. In wild groups many individuals are related, so that one observes ongoing parent–offspring and sibling relationships. Adults in the group probably first knew each other as infants – in fact nearly all relationships between individuals started when one or the other was born. This will be usually more general among females than males, since movement of males between troops has been recorded in several species, but movement of females is rarer, and when they move they tend to take infants with them. In contrast, caged groups are usually started by throwing together several strange adults. There is always a lot of fighting at the beginning, which with luck decreases gradually as the group becomes established. Since monkeys are known to learn rapidly and to have long memories, we can assume that these very different initial experiences of each other are going to have a lasting effect on individual relationships. It is reasonable to expect that in an adult female's interaction with an adult male there is some small influence of her memories of him when she was a juvenile handling him as an infant. This means that that particular male 'means' something rather different to her than does a similar adult male to a female who met him as a strange adult. This is a common-sense deduction, but there are a few observations which would tend to support it. There are many anecdotes illustrating the long memory of monkeys for individuals, but I found an incidental experiment of my own very impressive: I isolated females from a captive baboon group as they began to show menstrual cycles, that is when their infants were about six months old. They remained well out of sight and sound of the main group for seven or eight months, and were then returned, in many cases anaesthetized, to the colony area. The infants remained with the group, and were cared for by remaining adult females. As each female was carried into view of the colony, her infant, which had not seen her for more than half its life, began to make the specific 'lost baby' calls, and when she rejoined the group her infant clung to her and their relationship was resumed. The one exception was the first infant of the series, who had been completely adopted by a female whose own infant had died, and who did not desert his foster-mother. I am doubtful if human children would have remembered so well.

Caged groups that have lived together harmoniously for years may suddenly flare up into fighting as serious as that which occurred when they were first introduced. Possibly they never form a network of relationships like that of a natural group until the original stock is replaced by animals born into the group.

More responsible dealers prefer not to handle adult monkeys, especially females, because of their high mortality rate. When forming a new group, casualties are especially high among adult females. They are more aggressive towards strangers, especially other adult females; also, in the transition between capture and becoming established in a new captive group they are highly susceptible to diseases, many of which, like the dysenteries and tuberculosis, have been associated with stress. I think this suggests that the loss of her old network of relationships is extremely traumatic to an adult female, and establishing replacement attachments extremely difficult. It is hard to avoid the impression that some of these females simply 'die of a broken heart'. Less romantic evidence for this is provided by biochemical assay of adrenal function, discussed later in this chapter.

It is relatively easy to observe social interaction in large cages, and the same animals can be watched at the same time of day for the same length of time, eliminating many of the problems for the field worker. But what relation does the behaviour observed bear to what could be seen in the wild? If social behaviour is so grossly modified by the factors we have been discussing that the pattern seen is typical only of this environment, many of the results which could be obtained would be of little interest.

To try and assess the differences between social behaviour in wild and cage situations in one species, both the sort of differences and how large they were, I compared the data from the Ishasha baboons with observations on a caged group of baboons. In some ways the conditions of captivity were excellent: the caged baboons lived in an outdoor cement floor run, about 60 square metres, divided into five sections by mesh partitions with small doors. The cage was fairly typical of what anyone might try and keep a captive baboon group in, and it was about the same latitude as Ishasha, 320 kilometres to the east in Kampala, so one variable, the climate, was held constant for the two habitats. The caged group contained ten adult baboons and several infants. The composition was unlike that of the wild groups in that there was only one adult male, but this is typical of artificial groups because when strange adult males are placed together in a small cage they often fight and damage each other. The same observer collected both sets of data, which eliminated another variable. There were still enormous problems in making comparison between data collected in such very different places, and endless reasons for deciding that the comparison was scientifically invalid. But since there was only one study which gave information beyond the purely anecdotal (Kummer and Kurt, 1967), on what seemed to me a crucial point for primate studies if there is to be any transfer of ideas between cage and field studies, it seemed worthwhile making a series of enormous assumptions and risk the charge of scientific impurity and try to make the comparison. The biggest assumption was that an hour of field observation in good conditions was comparable to an hour of cage watching. The wild troops were much bigger, but were hardly ever all visible, and ten adults and some infants was taken as the average number of animals under observation at one time. There were three hundred hours of observation in each population, or roughly 3000 baboon-hours, on which the comparison was made.

The first point is that the same social behaviour patterns were seen in both places, communication patterns were identical. Comparison was therefore made on a quantitative basis, at three levels: the total frequency of interaction, the relative frequency of two main classes of interaction, friendly and approach–retreat, and the frequency and distribution of particular behaviour patterns in the interactions (Rowell, 1967).

Social interaction was nearly four times more frequent in the cage than in the wild. This difference was of course noticeable without any analysis: I was used to caged monkeys before the study started, and was impressed with how much time wild baboons spent just sitting around doing nothing much. They also spent more time foraging than the caged baboons spent feeding, and they also spent time moving from place to place, during which overt interaction was infrequent. In the wild, a quarter of the interactions were of the approach–retreat type, that is one of the animals involved avoided the other, while the remainder were friendly. In the cage the proportion of approach–retreat interactions was higher, at about a third of the total. Neither of these differences were evenly distributed among classes of animals: adult females interacted still more with each other in the cage than did other class combinations, and nearly half their interactions were of the approach–retreat type. These caged females had a markedly rigid hierarchy among themselves, whereas no hierarchy was observable among the wild baboons, and the hierarchical relationships of the other caged animals were less rigid than those of the females. We will return to this point later (*see* chapter 8). The behaviour patterns which were included in the approach–retreat interactions occurred in rather similar proportions in comparable age–sex classes in both populations. There were some differences which could be rather easily related to the physical differences in the environments – for example chasing was less frequent in the cage, which was hardly surprising considering that a full chase in the wild would be run over a hundred metres. Supplanting, on the other hand, was a more frequent component in the cage, again not surprising considering the relatively few desirable positions available in that environment. On the whole, however, approach–retreat behaviour seemed to be rather constant in behavioural detail, and to vary in quantity rather than quality.

Friendly behaviour showed more differences. Again patterns involving change in relative position were probably affected by the properties of the environment: both associating and following were recorded more frequently in the wild, but since the caged animals could not be more than fifteen metres apart anyway, these patterns seemed hardly necessary in the cage and were indeed less frequent. Grooming, and the gestures used to solicit it, were more frequent in the wild. The obvious response to that observation is to point out that the wild is an itchier place, full of irritating grass seeds, ticks, and biting flies, but in fact there is no evidence that grooming *other* animals is affected by such factors, and for birds, it has been demonstrated that mutual grooming, as opposed to self cleaning, is not affected (Sparkes, 1964). Other friendly gestures were more frequent in the cage: lipsmacking, putting hand or nose to genitals or elsewhere on the body, and embracing. These are all gestures of appeasement, reassurance, conciliation. In an earlier study of caged rhesus monkeys (Rowell and Hinde, 1963) in which we alarmed the groups very slightly, we found that in this stressful situation mutual grooming was greatly reduced, while lipsmacking, one of the reassurance gestures, increased. The pattern of differences in friendly behaviour, then, was consistent with the suggestion that the cage environment was more stressful than the wild.

The study of the captive group began within a few months of the baboons being put together, and most studies of captive groups have been made on recently established groups. Yet, as was suggested earlier, monkeys which first meet as adults may never establish the sort of network of relationships usual in the wild. Two studies that have been made on long-established zoo colonies give indications that these too may present problems for students looking for 'normal' social behaviour. Kummer and Kurt (1967) were able to compare an old-established hamadryas baboon group in Zürich zoo with the wild groups they observed in Ethiopia. Kummer had studied the same group when the adults were still the original wild-caught stock, and then four years later when many of the juveniles seen in the first period, which had been born in the zoo colony, had become adult. As with the Ugandan baboons, the captive hamadryas showed more social interaction than did the wild animals, and a higher proportion of interactions included approach/retreat patterns. Again, grooming was much more frequent in the wild than in captivity. Kummer and Kurt, however, also found differences in the communication patterns used in the two environments: two patterns were found in the field and not in the cage, and nine were found in the cage but not in the wild. (Some of these 'new' behaviour patterns, however, are known in other sorts of baboon, and since the exact origin of the captive hamadryas was unknown, it is possible that the difference already existed in the wild: where hamadryas and cynocephalus baboons overlap in range, they are said to interbreed freely, so that wild hamadryas stocks must have varying degrees of 'purity'.) The most interesting differences were related to the herding behaviour of the adult males. A wild hamadryas male threatens a female who strays and if she does not return catches and 'punishes' her with a bite in the back of the neck, whereupon she screams and follows him back to the group. The original adult male performed this herding fully, and led his females about the small enclosure in the usual way. The females, however, continued to scream after being herded until they had performed grooming, or intention grooming movements directed at the male's mantle, behaviour that was absent in the wild. The next generation of adult males performed the herding infrequently and imperfectly. In a unit led by one of these young males one of the original old females chased and punished quarrelling females in the way the adult male would have done in the wild (again behaving more like a cynocephalus baboon). Thus differences developed in the only set of behaviour patterns which distinguish the hamadryas from other baboons – patterns which must have been relatively recently developed. Herding and following seem to be inappropriate to an environment of about 350 square metres; it will be remembered that differences in frequency of behaviour used to maintain appropriate distance were seen in wild and caged cynocephalus baboons: following and chasing were less frequent in the cage, a parallel change to the one now described.

A rather similar comparison was made by Klein and Klein (1971) of a long-established spider monkey group at San Francisco zoo with a wild population. A single aged female was probably the last survivor of the original imported animals, and was the only one in the captive group which performed the full greeting rituals normal in interactions Klein observed in the field.

These observations are extremely important, and it is very much to be hoped that they will be confirmed and extended to other species, because they are the only real evidence that social behaviour may be affected by environmental pressures in content as well as in the relative frequency of components.

I have several times used the idea of 'stress' in comparing wild and caged monkeys, and it seems that captivity may be described as more stressful than life in the wild. Before we can use stress as an explanation for everything that goes wrong in a caged group from excessive fighting through epidemics of dysentery to failure to breed, we must be able to define it adequately.

At one level, stress may be defined in terms of the function of the adrenal glands. In response to stimulation by the pituitary hormone ACTH, part of the adrenal gland produces adrenocortical steroids. In a continuing stressful situation the adrenals enlarge and become capable of a sustained high output of these steroids. Such a modification is seen, for example, in overcrowded mice. Sassenrath (1970) used the technique of injecting ACTH into monkeys and measuring the adrenocortical steroids excreted in the urine to measure the responsiveness of monkeys' adrenals. She studied groups of five rhesus monkeys held together in small cages ($1\frac{1}{4}$m × $1\frac{1}{2}$m). She first carried out extensive observations on each group to determine its pattern of social interaction, then each monkey was placed alone in a metabolism cage, so that its urine could be collected for analysis. She found that a monkey's ACTH response was closely correlated with the amount of fear and avoidance behaviour it showed in the group and was moderately well inversely correlated with rank – alpha (i.e. top-ranking) animals excreted less adrenocortical steroids. After several weeks in isolation the adrenal responsiveness of the low-ranking animals dropped towards that of alpha animals, or animals held in permanent isolation. Responsiveness also dropped when the home cage situation changed: both when the alpha male was removed and when a female formed a consort pair with the alpha male. Females showed more fear behaviour, and had consistently higher adrenal responsiveness, than males of equal or lower rank. This finding seems comparable to the observation discussed earlier that caged female baboons showed more differences from wild females than did other classes, and formed themselves into especially rigid hierarchies.

Sassenrath's elegant technique allows adrenal activity to be monitored without killing the animals, but it still requires them to be removed from the social situation; as yet there is no way of directly recording adrenal activity in group-living animals, so that we are forced back on to behavioural measures of stress in continuing situations. Definitions of stress in behavioural terms are probably best left deliberately vague at present, though Sassenrath's correlations between adrenal responsiveness and a few well-defined behaviour patterns in one situation hold out hope of greater precision. A general definition of a stressful situation in operational terms is one which the animals try to avoid, and in which normal maintenance activities are disrupted.

We have so far compared wild and captive monkeys as if the latter were a homogeneous class, and have in fact referred only to groups held in relatively spacious quarters. Probably far more captive monkeys live in the more restricted conditions observed by Sassenrath, however, and studies are made on social behaviour in very small cages indeed. Some interesting differences emerge when Sassenrath's results are compared with a study of rhesus monkey groups of about the same size, but in runs more than twenty-five times as large (in terms of cubic space) (Rowell and Hinde, 1963). These monkeys were mildly stressed by offering food, by staring at them, and by approaching them wearing a mask and bizarre clothes. They showed stress according to the operational definition given above, by reducing the time spent foraging, grooming, sitting still, and in sexual behaviour, by avoiding the stimulus, and showing signs of fear of it, and by increases in urination, defecation, and other behavioural signs of uneasiness. Each monkey was tested in the home cage, alone, after six hours' isolation, and also in the group. The effect of isolation was greatly to increase the effects of the other stressors: isolation itself was a highly stressful experience. In contrast, the monkeys housed together in very small cages were more stressed than when isolated. I have frequently made the point that companions are an essential part of the environment for extremely gregarious monkeys, and this will be demonstated again in later chapters. Yet even this is not a valid generalization over the whole range of environments in which these hardy animals will survive.

5 Communication

Social behaviour is a matter of communication. Communication occurs when any signal given out by one animal is used by another to predict the behaviour either of the first animal, or of something else in their environment; we can only tell that communication has occurred when, after having perceived the signal, the second animal modifies its own behaviour accordingly.

Some signals are made incidentally, for example if a monkey starts to eat, the behaviour of an observing monkey may well be modified. Continual signals are given out by the colour and patterning of the fur and the skin about species, age and sex of a monkey. At the other extreme are gestures which appear to be made specifically to convey information, while in between are a series of movements and postures which provide highly important information for the social context, but probably have some other function as well.

Communication by monkeys is not qualitatively different from that of other animals, and the same principles apply to them. There is a complete break between people and other primates in this area, with the development of a verbal language capable of communicating about ideas, rather than objects or events in the immediate surroundings. Surprisingly, there is no evidence of any other primates showing even rudimentary forms of this type of language, although some recent experiments in teaching chimpanzees have indicated that they have some capacity for handling abstract concepts. People retain, in parallel to their verbal language, much of their basic primate communication repertoire of posture, gesture, and vocal quality (sometimes called 'non-verbal para-communication'), although some of these signals seem to have lost the precision of meaning their equivalents have in monkeys, perhaps because there is the possibility of correcting error using speech. One can usually identify emotions in pictures of monkey gestures or expressions if one knows the species, but those of our own species are surprisingly difficult – try, for example, the illustrations in Darwin's *Expression of the Emotions in Man and Animals* (1872).

Because communication is a very complex process, we can usefully approach it in several ways. It is convenient to discuss the sensory modalities separately, although communication is usually carried on through several modalities at the same time. Each modality has its own advantages and problems for communication, depending on the physical properties used. We can discuss the processes by which a given signal comes to be made, and the function of the communication system of which it forms a part in the biology of the animal.

Visual communication

First, let us look at monkeys. I choose visual communication first because it is probably the most important to monkeys and apes, and certainly the best described and understood. Primate eyes (and their associated brain mechanisms), as discussed earlier, have exceptionally good acuity, capable of perceiving very small movements and small differences in shapes. Visual communication has the advantage of instantaneous transmission, and several complex signals can be sent and received simultaneously. Visual communication is less good over long distances, and since light does not go around corners, the effective distance for visual signalling in the typical dense vegetation of monkeys' habitats may be rather short. It is possible for visual signals to be directed fairly exactly at one particular individual, and it is also possible to 'refuse' a signal, since the receptor organs, the eyes, are highly directional. 'Looking away' can be itself a visual signal: since a direct stare at another's face is a threat for all monkeys, looking away indicates non-aggressiveness – either mild fear or conciliation – depending on whether the animal looking away is lower or higher ranking that the other. (Remember this when looking at caged monkeys and try not to stare directly at them. It is also true of our own species, which is why children are taught that 'staring is rude' – rudeness is a form of aggression.) Altogether, visual communication is highly satisfactory for the processes of social interaction in a fairly compact group.

Postures

A great deal of information is conveyed by the general posture of a monkey as it sits or walks from place to place, by the shape of the vertebral column and the angles of the limb joints, especially the proximal joints, shoulder and elbow, hip and knee – the distal joints of wrist and hand have more limited use here because little variation in position is possible for them while being used to walk or forage. The monkey which lacks confidence is tense – it sits bolt upright, its head drawn into the shoulders, its limbs flexed tightly to the body, hands neatly down between the knees touching the ground. It walks with its back rather arched, knees and elbows bent as if it expected to have to jump to safety at any moment. A confident monkey sprawls comfortably when it sits, back slouched, legs stretched. Baboons commonly sit with wrists resting on knees, hands dangling free. When they walk, confident monkeys use a relaxed, rather rolling gait. Each species has its own variations on these general characters, especially noticeable in tail positions. For example the confident rhesus monkey carries its tail hanging rather loosely, one lacking confidence carries it stiffly out behind, while most high-ranking adult males carry their tails curled over the back. In baboons, by contrast, a vertical tail indicates fear, an unsure animal carries its tail rather stiffly and held slightly up, while the tail of a confident baboon hangs loosely after the first few stiffly fused vertebrae. An adult male baboon has a rather stiff-legged swagger and his tail swings jauntily from side to side like a Scotsman's kilt. Baboon and macaque tails are mainly used for signalling (several species of macaques have little or no tail showing that modern members of the group do not need one in locomotion). In species where the tail is still a long and muscular organ used for balancing, its carriage provides less information, though the normal carriage varies from species to species, and even from region to region (Langurs, *Presbytis entellus*, of northern India carry their tails up and arched forward over the back, while the same species in the south carry them up and then looping backwards).

Even where the tail is used mainly for communication, some common sense is required when interpreting its carriage. For example the tensed tail posture which indicates anxiety in baboons is also used to hold the tail out of the wet when walking through long grass after rain; a mother of a young infant may hold her tail vertical not in fear but to help her infant balance on her back; and the tail may also be held vertical while its owner is being groomed in the tail region.

Some of these postural points are illustrated in Figure 10. Using such cues, it is fairly easy to make a rough assessment of relative rank in a caged group before you see any overt interaction. The inconfident postures, though typical of low-ranking caged animals, would very rarely be seen in such an extreme form in the wild – perhaps immediately after a fight.

Postures will also indicate the level of arousal of monkeys – alert raised head positions and high tonus in the skeletal muscles indicate arousal. The skin muscles, which control hair positions also indicate excitement: monkeys' hair stands on end very readily. The hair of the head, neck and shoulders is especially mobile, and in many species a mane of longer hair in this area emphasizes the change in position.

Gestures

In understanding gestures, the most helpful concept is that of the *intention movement*, which was first formally characterized by Daanje (1950). The theory behind this concept is concerned with the motivation of the animal. If the animal is sitting in one place, and is about to move to another, its motivation to move changes along a continuum from zero to a level high enough to make it stand up and go. The change may be very rapid, but is more usually rather gradual, and the motivation may fluctuate during its rise to the level at which the animal moves. Before the animal is sufficiently motivated to make the complete

a Confident walk of adult male rhesus

b (Striding) confident walk of adult male baboon

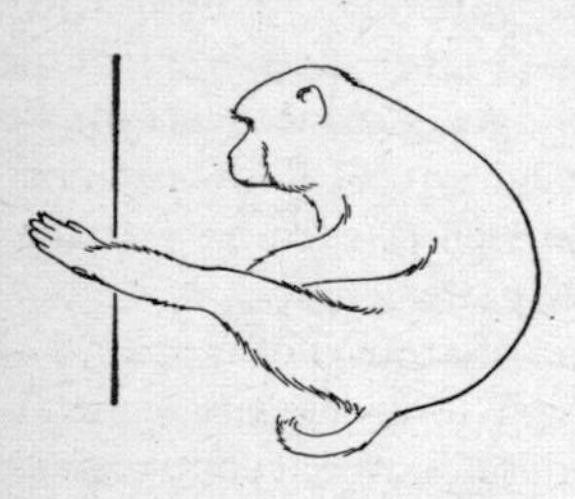

c Relaxed sitting position (feet up) of adult male rhesus

d Relaxed sitting position of adult male baboon

e Cautious walk of low-ranking female rhesus

f Cautious walk of sub-adult male baboon as he passes an adult male

g Sitting position of a low-ranking caged baboon female

h Cautious sitting position of low-ranking caged female rhesus

i Adult female (clutching infant) chases a higher-ranking animal which has just attacked her

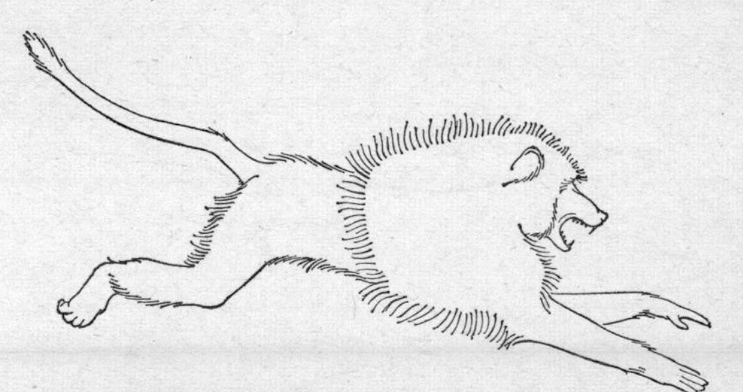

j Sub-adult male baboon chases a higher-ranking male returning an attack

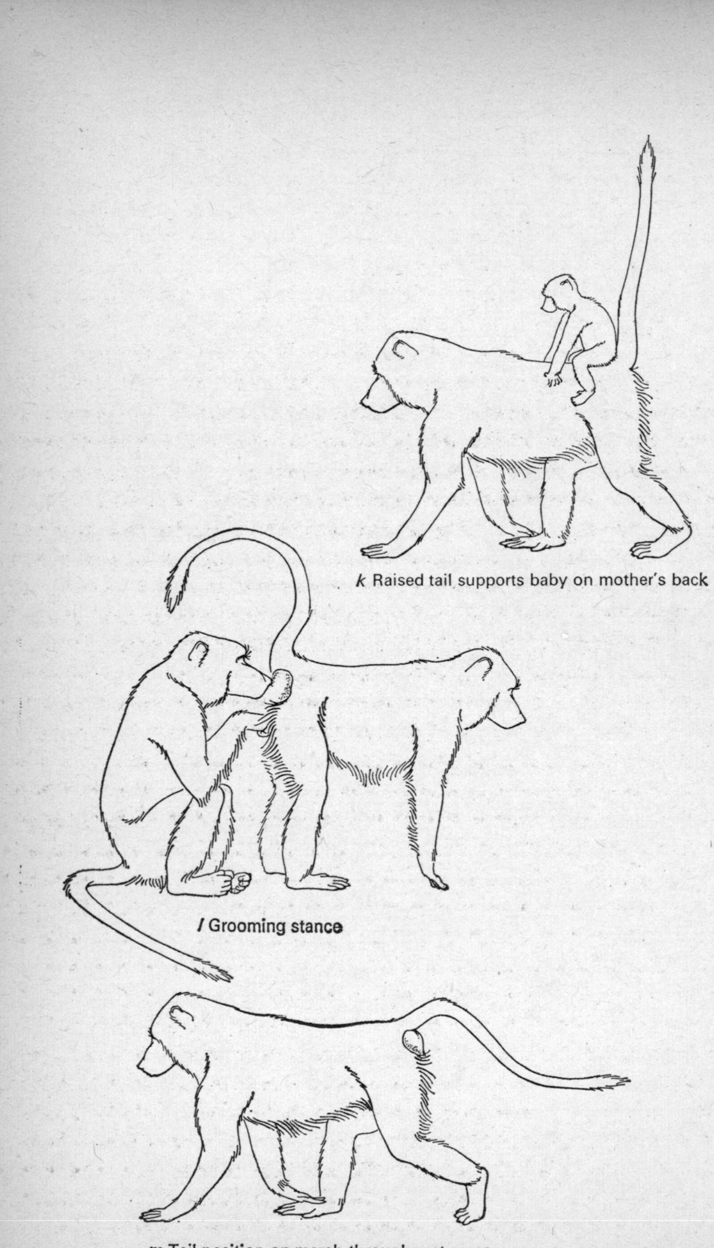

k Raised tail supports baby on mother's back

l Grooming stance

m Tail position on march through wet grass

n Tail carriage may vary locally

Figure 10 Communication by body postures. Note the line of the back, the angle of the limbs and the set of the tail in *a* to *h*. In *i* to *m*, some of the many situations in which one would record the 'tail-up' position in baboons. In none does it have the some meaning as 'tail-up' in rhesus macaques. In *n* the langur on the left-hand side is from northern India and the one on the right-hand side from eastern India.

move, it may make the first postural changes associated with starting to move, and the higher the motivation, the more complete such preliminary gestures will be. Thus a seated monkey will look in the direction of the water spout several times, and perhaps make a lurch forward – changing the centre of gravity preparatory to standing up – before finally making the larger lunge which will actually bring it to its feet. To the human observer or another monkey it is clear that that monkey is about to go and drink for some time before he actually moves.

An animal never wants to do only one thing at any given moment. Besides being a little thirsty, our example is probably slightly hungry; he has a slight itch where the fur on his leg is disarrayed, which needs scratching, while the place between the shoulder blades that he can't reach also needs grooming, and that would require soliciting grooming from a friend. He is comfortable where he is sitting, and has considerable inertia to stay where he is in spite of all these small things, but a higher ranking animal has been making intention movements of moving to the same shelf, and he may have to move out of his way anyway; there is also the receptive female he mated with earlier in the day who has moved out of sight and could perhaps with advantage be followed. If any of these other stimuli becomes more urgent, the moves towards the water may never be followed through; they will, however, have been noticed by the juvenile who happens to be sitting near the water, and will be supplanted should he move. Sometimes it is possible to do two things at once: If the higher ranking animal did move to supplant our example, he might well, at the same time as avoiding, go and get his drink. It would then be difficult to say whether his intention movements had indicated intention to drink or to avoid, and they would look no different if they had indicated both simultaneously. Often, however, two activities are incompatible – perhaps the food and water are in opposite directions from where the animal is sitting. In this case he might make an intention movement of rising, but the directional component is no longer clear – it may be made in the resultant direction, as if the animal was acted on by two forces. In 'attempting' to combine two rather different intention movements for two equally possible moves, the animal may produce an entirely novel movement.

The best series of examples of this is usually to be seen in fighting and threat behaviour. Aggressive and fearful behaviour may be referred to together as 'agonistic' behaviour, which is a useful descriptive category. Agonistic behaviour is composed of two very different sets of activities – one is moving rapidly towards the object and biting it, one is moving as rapidly as possible away. A monkey is rarely purely aggressive, without being a little fearful as well. A fearful monkey usually has some reason for not fleeing, be it inertia, some more positive reason for staying, or simply that he is confined and cannot escape. Threat and submissive gestures are thus seen as part of a continuum, and can be largely described in terms of combined or alternating intention movements of rushing forward and biting or fleeing, together with indicators of high level of excitement. An adult male baboon or rhesus monkey making a completely confident attack because his victim has not seen him coming makes a fast silent rush, opening his mouth as he goes ready to bite on contact. Less confident attacks are usually accompanied by noises which we shall consider later. The forward movement is somewhat slower and more controlled, so that it can be reversed. The mane is on end, the eyes wide, and the mouth open in preparation for biting – but not very wide, because the bite is not going to occur, the lips are relaxed and the mouth corners slightly drawn forward, and only the tips of the canines shown, the rest of the teeth are covered. Such an attack may be converted to a full attack if its object starts to flee, but it normally ends in a skidding halt. As the attack becomes less highly motivated, the rush is reduced to a lunge, then to a single blow on the ground with the hand, and then to a bob of the head. These threat gestures may be incomplete (as attacks) not because the threatener is frightened, but because his motivation is low, or his inertia is high. For example baboons will make hit-ground threats towards small juveniles which disturb them while playing. They are not frightened of these juveniles, but also it is not worth while to attack them more completely.

In the head-bob threat the balance between attack and flight may be beautifully demonstrated. The head makes two movements on the bob, one down and forward, the other up and back. The down-and-forward movement is the beginning of getting up and moving forward and the head then returns up and back to repeat the gesture. But the up-and-back movement is the beginning of getting up, turning around, and moving away. A monkey with a greater tendency to attack emphasizes the down-and-forward part of the head-bob, while a more fearful one emphasizes the up-and-back bit. By walking slowly towards a threatening female rhesus monkey one can obtain a demonstration of a complete range of intermediate head-bob threats, as with increasing proximity one becomes more and more alarming. The forward-emphasized head-bob is made with the mouth slightly open, corners forward. As the backward movement becomes more prominent the mouth closes, becomes compressed, and the corners begin to be pulled back. Most monkeys will stop head-bobbing and really move away at this point, but I once had a female rhesus with semi-paralysed hind legs, who because moving was difficult and hence inertia higher would stay longer, and demonstrate the further stages of the grin – as the corners of the mouth are drawn further back the front and then the side teeth are exposed, and finally the mouth opens, with all teeth bare, while the monkey begins to scream or squeak. This expression has been explained as deriving (Andrew, 1963b) from the 'facial defence' system – the expression made when sense organs are protected from a noxious stimulus, as when, say, sniffing ammonia fumes. It might also be derived from the tensing of the related muscles which turn the head – the conflict between turning the head to flee and keeping facing the fearful stimulus.

Once an intention movement has acquired meaning, or predictive value, and this is a matter of the evolution of the receiver rather than the animal making the movements, it may presumably come under selection pressure as a signal. A hypothesis about the processes involved here was put forward by Morris (1957). The broad selection pressures on communicative systems are that they should convey information efficiently ('efficiency' will have different criteria in different contexts) and that they should not thereby put the signalling animal at a disadvantage. The latter factor is most important when the animal relies for protection on being unobtrusive – being solitary and camouflaged, perhaps, but needing to communicate with others at least for reproduction. For the large social primates this is relatively unimportant, but can still be seen to operate in the communication of small cryptic monkeys like the talapoin. Thus when mildly alarmed, a group of caged baboons or chimpanzees will rush about and shriek, whereas talapoins will sit very very still, having at the first sign of danger moved apart, each to a different shelf.

Morris suggested that efficiency of communication would be improved if, in the first place, the number of possible signals were reduced. Instead of the infinitely variable continuous series we have been talking about, a species uses two or three of the possible gestures very much more frequently than the others – threats, for example are made at a few *typical intensities* and forms intermediate between these are rarely observed. Since there is usually no intrinsic communicative merit in any particular intensity, related species may select different typical intensities from the same series of gestures. Thus the whole family of the *Cercopithecinae* (baboons, macaques, guenons, mangabeys) have very similar total repertoires of behaviour, but gestures which are seen very frequently in one species are rarely recorded in another. The signal value may be further improved by making the gesture more conspicuous: three main stages in this process have been adduced. The emphasis on different parts of the movement may be changed, their time relationship may be altered, and finally there may be modifications of structure which emphasize the gesture. Classical examples of the last are the enlarged and colourful feathers in the wings of drakes which are shown off in some courtship gestures, or the plumes on the head of a heron which are shown when the bird makes a bowing greeting display on return to the nest. This process of improving the signal is called '*ritualization*'.

These processes are probably seen less clearly in the intra-group communication of primates than anywhere else. They are most convincing in well-armed species which maintain no personal relationships outside the breeding season such as sticklebacks, spiders and mantises. In a monkey group most visual communication occurs between individuals which see each other almost all the time. Signals are carrying a lot of information because they are maintaining a complex social organization, and the situation is such that considerable subtlety can be conveyed. An important factor is that, since individuals know each other personally, a signal will have different effects depending who sends it and who receives it – a juvenile would ignore, or even respond with teasing to a threat from a female of lower rank than his mother, but would flee screaming from the same gesture from a female of higher rank.

The agonistic communication set is the most strikingly continuous, but this important property has sometimes been obscured in the effort to describe it clearly. Where a list of gestures, like a simple dictionary, is available for a species, newcomers will tend to try and identify and match a gesture to each item on the list and then perhaps look no further. In many cases this will not matter, but a lot of information will be lost.

Less variability in the form of a gesture is found where immediate recognition of a rather precise signal is important, as in gestures used for conciliation or appeasement: among baboons lipsmacking, touching, offering and holding the hand, and embracing are relatively rigid in form (though embracing occurs in several distinct forms). Some of these gestures of conciliation have been described as developing from mother–infant interaction patterns – thus lipsmacking is derived from sucking, and in some greeting gestures towards adult females appears to be still directed towards the nipple; embracing is derived from the clinging and cradling of infant and mother (Anthoney, 1968; Ransom and Ransom, 1971). Presenting (orienting the rump to the face of another) is extremely variable on the other hand, perhaps reflecting its variety of use as a sexual invitation, a gesture of politeness, and an invitation to groom. Though presenting has often been stressed as an important gesture of conciliation – it has been supposed to reduce aggression by reminding the potential aggressor of sexual activities – in some quantitative analyses of interaction presenting behaves most like grooming invitations, which are also rather variable in baboons and macaques. Among guenons, by contrast, there is a notable lack of conciliation gestures, or at least they are used infrequently. Their grooming invitations are rather rigid in form, and may be preceded themselves by a series of intention movements; approaching another individual seems to be a more stressful enterprise for guenons than for macaques or baboons that may first conciliate from a distance.

Grooming invitations mainly consist of offering some area of the body to the groomer, while at the same time turning away the face. Each species typically offers a certain selection of areas at the start of a grooming bout much more often than others. A baboon usually offers hip or shoulder region ('lateral present'), while a rhesus monkey often offers the neck and chest, a Sykes' monkey (*Cercopithecus mitis*) the top of the head, a talapoin the back of the neck by lying down facing away from the groomer – to quote a few of the species with which I am most familiar.

Expressions, or facial gestures, have received special attention because of our own attention to them among ourselves. Although monkey expressions are obviously very closely related to our own, people are usually not very good at interpreting them without training. One widely misunderstood gesture is the 'grin' exposing teeth and gums, which, as we have seen expresses fear in Old World monkeys. A grin is commonly given by captive monkeys and interpreted as a welcoming smile by approaching people; alternatively, in different contexts, people refer the same expression to the canine signal system and describe it as 'baring the teeth', implying threat.

A detailed comparative study of facial expressions of monkeys and apes has enabled van Hoof (1967, 1971) to suggest origins and evolutionary paths which go beyond the facial defence system to a more complex system appropriate to the complexity of the material. He has shown how the grin (the silent bared-teeth face) has changed from a simple expression of fear to acquire more sophisticated connotations of conciliation and greeting (smiles) at least three times in different primate evolutionary lines, where social organization also became more sophisticated. Smiles are found in the mandrill (*Papio sphinx*), in the Celebes ape (*Cynopithecus niger*) in the chimpanzee and in ourselves. Laughter, on the other hand, which is often taken to be a more intense form of smiling, is closely related to the play expression of monkeys (the relaxed open-mouth face of van Hoof), and the convergence of laughter and smiling a recent human evolution which is only faintly presaged in the chimpanzee.

For the most part the gestures are used by monkeys in the same sort of 'sets' that we use ourselves, and so it is fairly easy to ascribe a meaning to each movement. Some are more baffling, however, because they seem to be used in several quite different situations. A good example (from Dr I. Bernstein) is a facial gesture of the pigtail (*Macaca nemestrina*) in which the chin is jutted forward and up, and the lips are pursed towards the other monkey (*see* Figure 11). It happens before an attack, before copulation, and when a mother retrieves her infant, for example. The only thing in common with these situations is that two animals become closer together, and the signal seems to be 'we are about to get closer' without any indication of what will happen next. Classification of signals into distance increasing, distance maintaining, and distance reducing has been used successfully for vocal signals also (Marler, 1968), so although by human standards it seems an almost meaningless way to classify what we want to communicate, it may be more generally applicable to signals of other primates.

Figure 11 Facial gesture of the pigtail monkey. Notice the forward-jutting chin and pursed lips.

Vocal communication

Next, we should listen to monkeys. Since noise goes round corners, it is an excellent medium for longer range communication where visibility is poor. It provides general information, which monkeys cannot shut out as they can exclude a visual signal so it can be used to attract attention, but cannot be used for exclusive communication between two animals. Sound signals are on the whole simpler than visual, only one can be made or received at a time.

Investigation of noises began with the development of tape recorders, and a device called a sound spectrograph, which will take about three seconds of recorded sound and make a picture of it by plotting the intensity of sound of each frequency in the audible range against time. Such a recording is not, of course, selective, as are ones' own ears, which listen to interesting sounds and ignore background noise, and so spectrographs include with impartiality the passing aeroplane, the bird singing, the slam of the cage door. In Figure 12, the monkey noise has been extracted from the background by tracing it, and somewhat simplified so that its basic characteristics can be seen. It is nowadays possible to discuss noises, using these aids, whereas previously the only method was to try and transliterate sounds – a hopeless attempt since letter combinations are said in different ways by different people even when they are ostensibly speaking the same language.

It is convenient to consider two sorts of noises separately, those which are made with gestures, and those which are made without.

The first group of noises are used in situations in which the animals involved can see each other, and seem to function mainly as emphasis, or attention direction, for the gestures with which they are associated. In most cases the gestures may occur without the noises, but not vice versa. Noises of this group are typically 'harsh'– that is they are relatively unstructured, with energy transmitted at a wide range of wavelengths.

The crude noise produced by the vocal cords is modified by

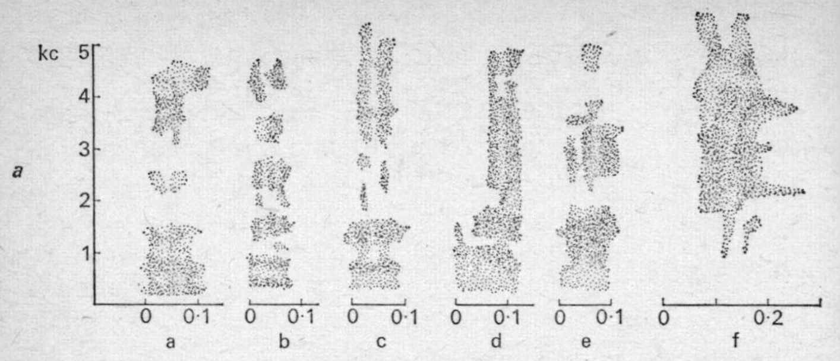

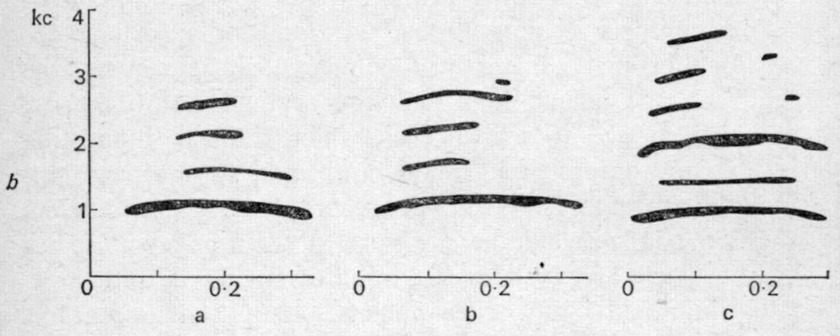

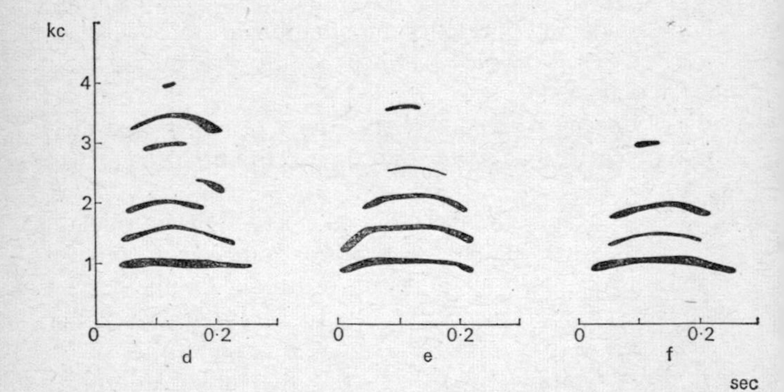

Figure 12 Spectrographs of some rhesus monkey noises
a These are all harsh noises
b These are all clear calls
Source Rowell and Hinde, 1962

the pattern of breathing, and by the resonance chambers of the mouth, throat, and nose. As Andrew (1963a) has pointed out, the information such a sound carries is mostly about the positions of these resonating chambers. Bearing this in mind, a set of noises such as the agonistic noises of a rhesus monkey, should not be thought of as a series of 'words'. They can be more informatively examined by considering their qualities separately, and relating them to what the resonance chambers are doing (Rowell, 1962).

Although the noises have a wide frequency spectrum the distribution of energy is not even, so that they give the impression of differences in *shrillness*. These differences are due to the position of the mouth, in the same way that vowels are differentiated in human speech. In the series u-o-a-e-i the lips are drawn back and successively higher harmonics of the base tone become emphasized, so that 'i' is a shriller sound than 'a' even though the sound produced by the larynx is not changed. We have just seen how the position of a monkey's lips depends on how frightened it is – the more confident, the farther forward its lips are protruded, the more fearful, the more the corners of the mouth are drawn back. These changes of expression occur whether or not a noise is made, but they mean that a frightened monkey will produce a shriller noise than a more confident one, from a given laryngeal sound.

The more vigorously air is expelled, the louder the noise produced. Also, the lower frequencies need more force, so that in a quiet noise the higher frequencies are relatively more audible, as in the difference between whispering and ordinary speech. When air is expelled gently, the back of the tongue and throat can vibrate at a very low frequency, giving the effect of a guttural 'r'. The force of expulsion is mainly governed by the position and movement of the head and neck. When the head is jerked down and forwards at the enemy in threat, air is expelled very forcefully, and the noise produced is a loud bark. In less aggressive threat the head is jerked up and back, and in alarm it is held up and back into the shoulders, the chin in to the neck. In this position it is almost impossible to force air out hard, and so the back of the throat and tongue vibrate. The result is a rolled 'r' effect heard in the growl and shrill-bark group of noises made by alarmed animals.

The length of the noise depends on the type of breathing at the time, and this changes with what the animal is doing. When sitting still breathing is fairly shallow, and a noise made on such a breath is short. As activity increases breathing gets deeper, and noises longer – thus in mild threat a rhesus monkey barks, while when more excited he may roar. When the animals are over-exerted and get out of breath, noises tend to become shorter again: an excited and rather frightened monkey screeches, an attacked animal makes shorter and shriller screams, while a defeated and exhausted animal makes short squeaks.

When a rhesus monkey becomes very frightened or 'excited' the noises it makes take on a scream quality, caused by strong constriction of the larynx which almost eliminates the lowest 3-4 kilocycles of the frequency range. The context suggests a function of the sympathetic system.

Another character of this group of noises in the rhesus is a syllabic quality, produced by rapid changes in the position of the diaphragm which break up the noise on one breath into several shorter units, rather as in laughing. Like laughing, this quality appears when the animal is emphasizing social bonds. In the monkey it occurs in threat when the monkey needs support of the rest of the group, or in the scream of an attacked monkey which needs rescuing by a higher ranking animal.

All these modifications of a basic harsh noise are themselves variable, in that they may be weakly or more strongly present, and they occur in combinations. This produces an infinitely variable system of vocal communication, capable of signalling very subtly about agonistic situations. The system is complete in itself in that it is possible to follow an interaction which is out of sight by listening to the noises. The changes in the noises are produced, however, by movements, mainly of the head, which are not primarily concerned with sound production, and which happen even if no noise is being made. To understand the derivation of the noises, it is necessary to consider the whole animal, because the noises are not at all emancipated from the gestures used in the situation.

Noises which are made without gestures (and often they are made with so little movement that it is very difficult to tell which animal made them) are usually much more structured than those we have just been discussing. That is, their energy, far from being spread rather evenly over a wide frequency spectrum, is concentrated into a series of narrow bands. They could be described as clear, or even tuneful, rather than harsh. Noises of this group are used for two rather different purposes, for which different physical characteristics are required: these are maintaining contact within troops, and maintaining contact between troops.

Monkeys of a group are usually not able to maintain continual visual contact because they live in dense vegetation. They can maintain group cohesion by making occasional calls which indicate the general whereabouts of each individual. These calls become more frequent in situations where they are likely to lose contact – when the group starts to move, when some of the group locate a new food source, or when any other important movement in the environment occurs which might necessitate troop movement. To translate to the cage situation in which these calls have been recorded, they are made when a monkey moves after having been seated for a long time, or when she sees someone else move; when food is brought towards the cage, or when the caretaker comes in sight, or even when the sun appears suddenly from behind the clouds. These calls could also make the individual monkeys more vulnerable to a predator, but this possibility is reduced by the structure of the noises used. Because they are short, rather indeterminate in beginning and end – they fade in and out – and because they use rather few frequencies, they are extremely difficult to locate precisely (this difficulty is heightened by the lack of movement when they are made). The significance of these properties of animal sounds was worked out on bird songs by Marler (1957). It is probable, however, that each animal's voice can be recognized by other members of the group – that is, the sound conveys a general message 'monkeys around here' and also a message only available to the animals to which it is directed 'experienced old female X is somewhere over there'. (This possibility has not yet been explored, but slight variations in the structure of the calls which can be seen on spectrograms indicate the likelihood.) The difficulty of localizing the sound is obviously more important for small, cryptic species (that is those whose appearance and behaviour hide them) than for larger animals for whom predation is less of a problem because they can defend themselves from most predators. The tiny talapoin has an extremely difficult call to locate, that of the rhesus is rather less so, while the deep contact grunt of the baboon is relatively easy for our ears to pinpoint.

Contact calls have been described only in species which move about in relatively large, compact groups, like the baboons, the macaques, and the talapoin; in the other guenons, whose troops tend to break up into small foraging subgroups, contact calls have not been recorded. It is possible that they exist but are too quiet to be heard in the field: contact calls are much rarer in a bare cage where all animals can maintain visual contact, and perhaps in some species they are not made in captivity. They are most often, though not exclusively, made by adult females, which fits with our earlier observation that adult females form the nucleus of troops.

At least some species, and probably most, have calls for re-establishing contact once this has been lost. These are best known as infant-lost calls, but the baboon at least has a similar call used by lost adults. These calls are similar to the contact calls of the species but much longer, and consequently easier to locate. In the baboon and rhesus infants they may be combined with click series which again increase localization. These calls may be made by the lost animal or by those trying to find it – the lost infant calls, and when a baboon was removed from a group her adult daughter made lost calls from a high perch, apparently looking for her. These contact and lost calls serve to reduce distance between individuals, or at least to prevent it from increasing. In communication between troops, maintaining or increasing distance seems to be the important function. Most monkey groups overlap their home range with that of several other groups. They may have little overt interaction with the monkeys of these groups, but they are intensely interested in whatever they can see and hear of them: in a communicative sense each monkey belongs to the whole local population of his species as well as to his immediate group.

Each species has a highly distinctive noise which is used for inter-group communication. It is usually made by the adult male, and in several species the male has large resonating sacs in his throat which give the sound a booming, carrying quality (compare the colobus, the howler monkey and the gibbon). In contrast to most monkey noises, which are not easily distinguishable in closely related species, this noise is instantly recognizable and proclaims the presence of that species. Because they are to be readily identified, these noises are highly structured, either, like the contact calls, in terms of restricted energy bands, or they have a distinctive time course, like the long song of the colobus or the two-syllable bark of the baboon. Because they are structured, and because they are loud, they carry well over distance and are not easily lost in background noise. These noises are usually made after any disturbance, which might have caused change of position of groups, and because of this they are often described as alarm calls, which they are not. They are also given spontaneously, or at certain times of day, notably around daybreak; and they are very readily given in response to the same call from another group. Marler (1969) has pointed out that spontaneity is an essential characteristic of signals which maintain distance between animals, since they are not necessarily preceded by a change in the environment which would act as a stimulus for making them.

Alarm calls are made by all monkeys, not just adult male. In structure, they are related to the agonistic noises discussed earlier rather than to the contact calls – they are high-pitched, short noises with a wide frequency spectrum. Alarm calls of most species are rather similar, and monkeys respond to each other's alarm calls when they live in the same area. Unlike all the other noises we have considered so far they do not seem to carry information about who is making them: if tape recordings of noises are played to monkeys they will respond appropriately to alarm calls, whereas most other noises elicit only mild interest, since they are made by strangers and therefore of no direct concern.

Chemical communication

Scent has the great advantage that it can be left to transmit information while the animal that made the signal goes away and does something else. It is, however, a relatively slow form of communication, and probably incapable of transmitting very complex information or of changing content rapidly. The primate order includes both the mammals which use scent least of all and those in which scent marking behaviour is most elaborately developed, so that it is impossible to generalize about primates in this respect, except to point out that whereas among the carnivora (dogs, ferrets, etc.) or the insectivora (shrews, hedgehogs) the chemical senses are used to obtain all sorts of information from the general environment – to find food, to follow paths, and so on – the primates use their eyes for the most part for these activities and use their noses almost exclusively for social interaction.

The most highly elaborated olfactory communication of all is seen among the prosimians. The skin of lemurs is full of scent glands: there are sebaceous glands among the hair roots on the general body surface, and enlarged sweat glands especially on the face and lips. In addition they have special scent organs, large enough to be visible in the live animal. The ring-tailed lemur (*Lemur catta*), for example, has highly glandular skin on the scrotum or the vulva, a complex scent gland and spur on the wrists, and a large brachial gland just where arm and chest meet. Both sexes rub their genital glands against twigs; and males have two further scent-marking ceremonies. They palmar mark, holding a vertical twig and pulling at it in jerks, and most elaborate of all, they tail-wave. In this ceremony, the wrist glands are first rubbed together, and then each is rubbed on the brachial gland, probably getting scent from it on to the wrist spur. The long tail is then brought round and the tip thoroughly rubbed between the wrists, and finally the tail is arched over the back and its scented tip shaken in the face of another lemur. Another male may do the same ceremony in return or he may run away; a female usually cuffs a male who tail-waves at her.

We might suppose that each of these glandular areas would have a different signal function, but we have very little idea what sort of information is conveyed by the different secretions. In the brown lemur (*L. fulvus*), which includes a large number of rather different looking races or subspecies, Harrington (1971) has been able to show that the scent of the perineal gland is not subspecies specific, and that the animals recognize individual scents. This is a good start in developing a very difficult experimental technique, but it is reasonable to assume that such a complex battery of scent organs conveys many more sorts of information.

Olfactory behaviour is also obvious in New World monkeys. When two adult night monkeys (*Aotus trivirgatus*) meet, they sniff nose to nose, nose to anogenital area, and nose to armpit; there are scent glands in all these places. Tamarins (*Saguinus geoffroyi*) rub their anogenital regions on branches, using two quite distinct movements depending on how excited they are, while titis (*Callicebus moloch*) rub the chest, and presumably the brachial glands (Moynihan, 1964, 1970). In several New World monkeys the clitoris of the female is enlarged. Klein (1971) has shown that the long, cupped clitoris of the spider monkey is used for urine marking, perhaps mixing urine with glandular secretion; males will climb some distance to smell and lick urine deposited by females, or even just the place where a female has been sitting, and this behaviour is not just related to oestrus in the females but is directed to all adult females. This same species also has a greeting ceremony in which the partners embrace and each puts its nose to the brachial gland of the other. Urine marking is well known in male squirrel monkeys (*Saimiri sciurus*) – the monkey washes hands and feet in urine which is then spread as it walks (some of the prosimians mark in the same way) and they also urinate towards their opponent in the penile display, which presumably adds olfactory emphasis to this aggressive gesture.

In contrast to all this olfactory activity, the Old World monkeys and apes have until recently been assumed not to use olfactory communication. They have small noses, and no obvious specialized scent glands in the skin. Some behaviour, however, suggests that smells may be important after all. Baboons and chimpanzees will touch a stranger and then smell their hands; guenons put their nose to the mouth of another in a common greeting pattern. Often nose-to-mouth is directed at a feeding animal, but it is also usually directed to higher ranking and older animals. Rubbing the chin and chest on the substrate has been seen in male vervet monkeys and is possibly a way of scent marking a territory (Gartlan and Brain, 1968). Putting nose to genitals is fairly commonly recorded – in Sykes' monkeys male and female circle as they nose each other's genital region prior to copulation; male baboons put their noses to the swellings of oestrous females, or may insert a finger into the vagina and smell it carefully. Naso-genital investigation is not frequent in rhesus monkeys, but Michael and Keverne (1970) found that the vaginal secretions of a receptive female, if spread on the rump of a non-receptive one, will cause a male to treat her as if she were receptive.

Chemical communication is only beginning to be explored in animals in the last few years, and I expect our knowledge of it in primates will soon be greatly expanded. Since we are probably the least osmatic of all primates, experimental techniques have been slow to develop.

Tactile communication

Touch is probably of enormous social significance for all primates, but is almost impossible to study because the signal cannot be intercepted. The importance of tactile communication in our own species has only recently been recognized academically (Bowlby, 1958). The most frequent interaction pattern between monkeys is mutual grooming, and whereas this undoubtedly does have a function in skin care, its communication value is more obvious, if only because grooming is very unevenly distributed among members of a social unit. Some monkeys spend a lot of time huddled together in maximum skin contact. Kaufman and his co-workers have demonstrated that this behaviour has communicative function, in the sense of modifying behaviour. They compared bonnet and pigtail macaques, the former being a huddling species, the latter being relatively infrequently in contact. Pigtail monkeys, as their colony matures, develop strong matrilines – monkeys are friendly almost exclusively with their relations through their mothers. Bonnet monkeys live much more communally, and relations are treated not very differently from other members of the group. When a pigtail infant loses his mother, the only animal with whom he normally has much contact, he undergoes severe depression from which he recovers only slowly. A bonnet infant is literally taken to the bosom of his wider family, and after the first period of anxious searching shows no clinical symptoms of depression.

Some non-huddling species, such as the baboon and the chimpanzee, still use touch for greeting and reassurance, but instead of general body contact a formal embrace or brief touching of hands is used. Tactile communication also includes grabbing or holding fur, biting, tail-pulling, etc. In most cases such aggressive contact occurs amongst other agonistic behaviour, but among guenons it has perhaps been to some extent ritualized for communication of rank without aggression. A high-ranking vervet female will approach a lower ranking one and firmly grab her topknot and pull her head down, and hold it there for a moment; she may then go on and groom, but the first gesture looks uncomfortable, though the grabbed animal does not protest. Such behaviour is even more striking in the talapoin – a talapoin female will pick up the tail of another and bite it, and the other pulls away without noise or change of expression; the alpha female of my group will lean over and bite across the back of the female sitting next to her, again in what looks to be a firm enough way to be painful, and again the bitten female quietly waits till she is released and then moves slowly off. This is very unlike the usual energetic response of an attacked monkey, and suggests that these 'attacks' are conveying information rather differently. Talapoins will also make the topknot pulling gesture described above for the vervet, and then make the usual grooming movements of fur parting and cleaning very roughly. This can only be described as 'aggressive grooming', though the phrase would be a contradiction in terms for any other species. In this discussion it is necessary to use words like 'looks painful' – there is great difficulty in assessing either stimulus or response. It is possible to learn more about tactile communication from pet monkeys, or by hand-rearing infants, and I have found such relationships highly illuminating, although it cannot be stressed too often that monkeys make highly unsatisfactory pets for anyone except the totally dedicated. The problem of communicating verbally about tactile stimuli remains unsolved, however.

At the beginning of this chapter I pointed out that primate communication was not qualitatively different from that of other animals, until we come to human verbal language, so that it is possible to apply general zoological principles in the usual way. Monkeys are only approached by a few other animals however in the *quantity* of their communication – they spend a large part of their time in giving and receiving social signals. Their nervous system is so adapted to social life that they seem to need to give and receive signals in the same way that they need vitamins in their diet, in order to remain normal healthy animals. An isolated monkey will work for the sight of other animals in the same way that a hungry one will work for food. This being so, one should use the same humanity in considering an experiment involving isolation as one would use in designing one requiring food deprivation.

6 Reproductive Behaviour: Mating

Mating behaviour has been investigated from two main standpoints: It can be considered as an interaction between two individuals, in which case the object of research has been the factors controlling when and how much mating occurs relative to changes within those individuals; and it can also be considered as behaviour of the population, when the question is how far mating and more especially conception is synchronized, and what environmental factors might be affecting the seasonality of breeding. Interest in both these aspects was focused by Zuckerman, in his classic *The Social Life of Monkeys and Apes* (1932), in which he combined his own observations with ideas of primate behaviour then current to formulate a theory of primate social organization. He suggested that the societies of primates were held together primarily by sexual attraction between males and females, and that their high degree of permanence and resultant complexity of structure was made possible by year-round breeding, and the relatively long periods in which females were receptive. Although many of Zuckerman's assumptions have since been disproved, his synthesis still remains a stimulating cause of argument.

Breeding seasons

One of Zuckerman's mistaken assumptions was that all primates are much the same. He went himself to see whether primates in the wild had a breeding season or not, and sensibly chose some of the most easily available, the chacma baboons in the Cape province in South Africa. He found cycling, pregnant and lactating females in the same troop at the same time, and concluded that they had no breeding season, and this gave him confidence to disallow the many reports available to him from tropical naturalists of breeding seasons observed in other species. (Zuckerman may be forgiven for being tempted to generalize about the group since he had just made the very important discovery that the menstrual cycle is a general characteristic of the order.)

Although his data were rather sparse, Zuckerman was right about the chacma baboon, whose year round breeding has since been confirmed several times, and is the more surprising because the climate at the Cape included large seasonal variations in temperature and easily detected changes in day-length. At Ishasha, and at Gombe Stream, which are both close to the Equator, baboons also breed all the year around, and yet in Nairobi and at Amboseli, at almost the same latitude, births are much more frequent at the onset of the main rainy season (October to December) than at other times. Other species are still more exclusively born in a limited birth season, and throughout the primates a restricted birth season, or sometimes two in the year, is the rule, rather than the exception (Lancaster and Lee, 1965).

How breeding seasons are determined among primates is by no means clear. The problem is particularly fascinating among mammals with a long gestation and a long lactation, because it is not at all obvious which period is the critical one. In small birds or rodents it is obvious that the brief intense effort of reproduction should be concentrated during the season when most food is available. In the larger mammals, though, are metabolic requirements at their largest during pregnancy, early lactation, or perhaps the period of weaning the infant to solid food? These three periods occupy nearly the whole year. If the critical period – that is, the period in which selection pressure acts most strongly against animals that are incorrectly timed – is, say, early lactation, it is separated by perhaps half a year from whatever trigger stimuli begin the reproductive process. In teleological terms, the species must 'know' that for its babies to survive best they must be born in June, and therefore copulation must happen in December: conditions which identify December as the right time to mate have nothing to do with the conditions in June which determine the selection pressure.

In mammals and birds which live in high latitudes, change in daylength forms the trigger which synchronizes breeding. It is the change in daylength rather than its absolute length, which acts as the stimulus, and this change is perceived by the eye; the information passes, via the hypothalamus, to the pituitary gland which increases its secretion of gonadotrophic hormones which in turn stimulate the ovary or testis to secrete their own hormones, causing the females to come on heat and the males to court them. In animals with long gestation, decreasing daylength has this effect, while in those with short gestation, increasing daylength is the stimulus, so that both give birth in the spring. Macaques range into quite high latitudes, where daylength changes are obvious, and they give birth generally in the spring, having conceived nearly six months earlier in an autumn mating season. This fits with the idea of a decreasing-daylight trigger mechanism for the breeding season, but this has not been demonstrated experimentally even in the most northerly species, probably because when brought indoors in the laboratory macaques breed at any time of the year.

The nearer the Equator, of course, the less obvious the differences in daylength; in the tropics the obvious seasonal differentiation is into wet and dry rather than dark and light, and plant growth and fruiting seasons are also correlated with rainfall. It is still possible to use daylength changes to synchronize behaviour – the red locust does this at such low latitudes that the daylength changes only a few minutes on either side of the Equinox. It is also possible to respond directly to rainfall, as do some finches, or to the sight of the new growth of grass, as do others. The cues used by tropical primates are not understood. Breeding seasons are usually described in relation to rainy or dry seasons, and Vandenbergh and Vessey (1968) described colonies of rhesus monkeys near Puerto Rico which showed peaks of mating associated with rain and the new growth of leaves. The relationship between the two phenomena was not very precise, however, so that it is difficult to envisage a possible physiological mechanism of response (the rhesus is not native to the tropics). The problem is further complicated by the fact that the rainy seasons are themselves related to the calendar: at the Equator there are two rainy seasons a year, roughly at the Equinox. Further from the Equator one rain becomes relatively more important and the other diminishes and one dry interval becomes longer and the other shorter, until at the edges of the tropics there is only one rainy season a year. The rain may vary in its exact timing, and there is some evidence that the plants may time their new growth not just as a direct response to rainfall but partly in relation to the calendar – possibly responding again to daylength changes however slight. Thus they shoot relatively earlier in late rainy seasons. Monkeys generally have one birth peak even where there are two rainy seasons though a second smaller peak, six months out of phase with the main one was noted in rhesus monkeys in India (Prakash, 1962), and a captive colony of Sykes' monkeys, on the Equator, bred twice a year for a time (they conceived during the dry seasons, not in the wet seasons like the rhesus colonies mentioned above).

More directly relevant to our theme is the possibility of synchronizing breeding activity through social stimuli rather than by a general response to a change in the environment (the two mechanisms would most probably be complementary rather than exclusive). Evidence for such an effect is again due to Vandenbergh (1969). Following the birth season, rhesus colonies have a sexually quiescent period which lasts most of the summer, during which they moult, and sexual skin is pale, and even females which are not caring for young infants show little oestrous behaviour. The testes of the adult males become smaller during this period, and there is little spermatogenic activity (Sade, 1964). If such sexually inactive males, however, were exposed to an ovariectomized female treated with oestrogens to make her sexually receptive, they would copulate, groom more, and show increased testicular activity. There are no good data on the reverse process, but a suggestion that something similar might occur came from vaginal lavage data on one Sykes' monkey. She had had an infant some three months before and was still completely sexually inactive, as indicated by the few cells seen in the vaginal fluid. The first change in this picture was the appearance of sperm, indicating she had been mating, and a week or two *later* her vagina began to show signs of cyclical activity and increased oestrogen levels. Thus we have the possibility that relatively slight changes in behaviour in response to environmental stimuli might be escalated into a major change in group behaviour via mutual excitation. This could result in a much more precise synchronization of conceptions and births than if each individual separately responded to the environmental change, with varying sensitivity.

Synchrony of births can be advantageous in itself, regardless of which season they occur. On the Athi Plains in Kenya, wildebeeste all give birth during a very few weeks in the year, and during this time wildebeest calves are the main food of hyenas and other predators. Since only a limited number of predators can survive the rest of the year, the predators can take only a small fraction of the calves born in the middle of the season, but they take a much higher proportion of the stragglers born early or late. Among primates, however, heavy predation on infants has not been identified as a major controlling factor to date. There has been speculation that concentration of births at one time means that, with all adult females preoccupied with infants and none sexually receptive, fighting would be at a minimum when the infants were small, and that their chances of survival thus enhanced.

Studies of mating behaviour as an interaction pattern between individuals have nearly all been made in the laboratory, where breeding seasons tend to disappear and copulation can occur throughout the year. They have not, for the most part, considered the possibility of seasonal changes in the behaviour studied, although in most macaques, for example, complete copulation series are mostly confined to three or four months of the year in the wild. In a captive vervet group, I saw another possible arrangement: There was very little seasonal change in frequency of copulation and other sexual activity, but conceptions occurred only during part of the year, suggesting a direct physiological effect without behavioural intermediary.

Individual reproductive cycles

The chief variable which has been studied in relation to mating behaviour in the laboratory has been the female reproductive cycle, with almost all the emphasis on the menstrual cycle rather than pregnancy and lactation interval, although the latter states occupy much more of the life of a normal wild adult monkey. Many studies are actually of castrated animals given replacement sex hormones, but in most cases this procedure is aimed at reproducing the effects of the normal menstrual cycle under more rigorously controlled conditions. We should first, therefore, have an outline understanding of the female primate's reproductive cycles. The following account refers only to the Old World monkeys and apes, because data on New World monkeys and prosimians is fragmentary, and in some cases suggests that their cycles may be rather different from those of the better known Catarrhines.

Background physiology and anatomy

Menstrual cycles are usually taken to start on the first day of bleeding, and are about four or five weeks long. Ovulation occurs about the middle of the cycle, so that in the first half, the ovary contains a developing ovarian follicle, and after it has ruptured and released the egg, the follicle is re-formed into a corpus luteum, a secretory structure which lasts for about two weeks and then, if conception has not occurred, is resorbed. At this time uterine bleeding occurs, starting the next menstrual cycle. The life of the corpus luteum is rather constant in length, particularly if successive cycles of the same female are considered, so that there is relatively small variation in the length of the second, luteal, part of the cycle. The time from onset of menstruation to ovulation is much more variable, mostly in the direction of lengthening rather than shortening the cycle. Thus records of cycle length in populations of monkeys of several species all show a characteristic skewed-normal distribution with a long tail of unusually long cycles (*see* Figure 13). It seems that various factors can act to delay the maturation of the follicle during the first week of the cycle, but that once a certain stage of development has been reached the rest of the maturation process 'goes off' in a rather invariable period (roughly another week). The drug reserpine has been shown to lengthen cycles if administered in the first week, but has little effect later (Erikson,

Reynolds and de Feo, 1960). Various forms of stress have been shown to have the same effect, including transferring to a strange place, which may prolong the cycle to such an extent that the female is reported as becoming acyclic (compare Gilman and Gilbert, 1946). Social factors can have the same effect: in a baboon group nearly all the unusually long cycles occurred after the female had been moved, or had been beaten up by other females around the time of menstruation (Rowell, 1970a). Data on baboons, chimpanzees and rhesus monkeys give longer mean cycle lengths for group-living animals, either wild or captive, than for isolated females, suggesting that this effect of social stress may be rather general. The effect may be more important than might be expected because being beaten by other females does not occur randomly through the cycle but occurs most often around the time of menstruation in rhesus monkeys, and probably in baboons.

The growing ovarian follicle secretes large amounts of oestrogens, whereas the *corpus luteum* secretes mainly progesterones. The menstrual cycle as usually detected is a cycle of growth and breakdown in the lining walls of the uterus and vagina.

In some monkeys there is an area of naked skin surrounding the perineum which is also sensitive to gonadal hormones. Zuckerman called it 'sexual skin' because of its response to sex hormones, which includes changes in colour and oedema or swelling. The behaviour of this skin is not the same in all species which possess it, and considering its taxonomic distribution, it must have been evolved separately several times among the Catarrhines (*see* Figure 14). The infant rhesus monkey, for example, is furred around the perineum. In females, at adolescence, the skin between the legs, over the tail root, and down the posterio-lateral edge of the thighs begins to swell during menstrual cycles – it increases steadily through the cycle until a day or two before menstruation and then quite rapidly decreases. During these early, rather erratic, and infertile cycles the skin gradually reddens and becomes very sparsely haired. As she enters her second breeding season the young female stops swelling, and begins to show only rather variable fluctuations

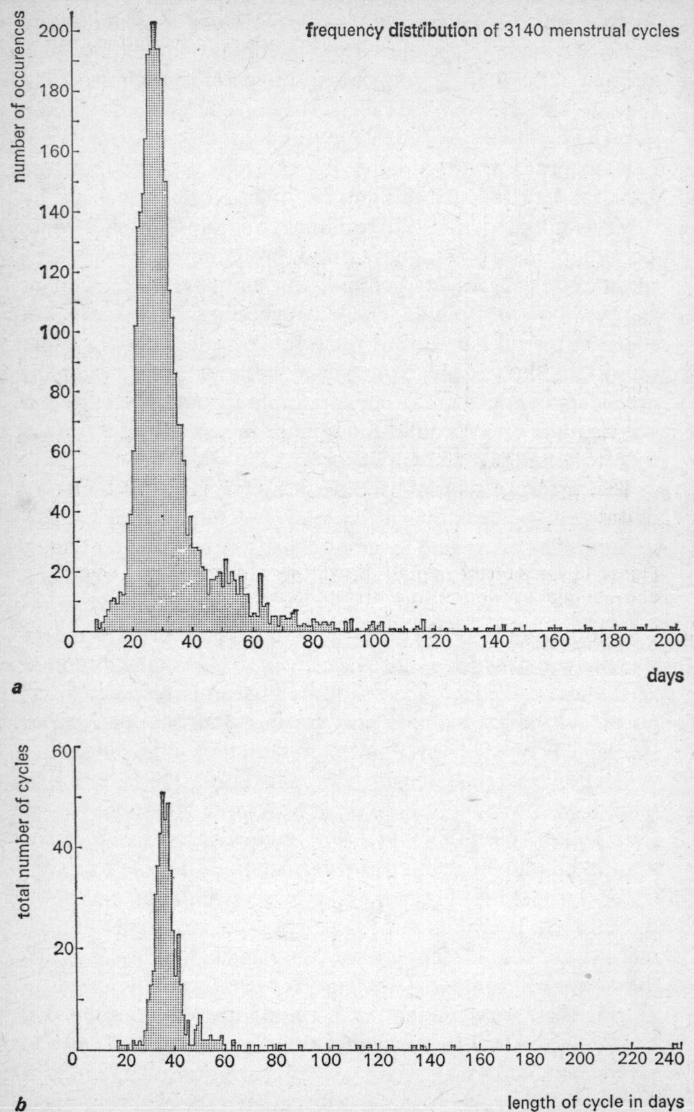

frequency distribution of 3140 menstrual cycles
number of occurences
200
180
160
140
120
100
80
60
40
20
0
20
40
60
80
100
120
140
160
180
200
days
a
total number of cycles
60
40
20
0
20
40
60
80
100
120
140
160
180
200
220
240
b
length of cycle in days

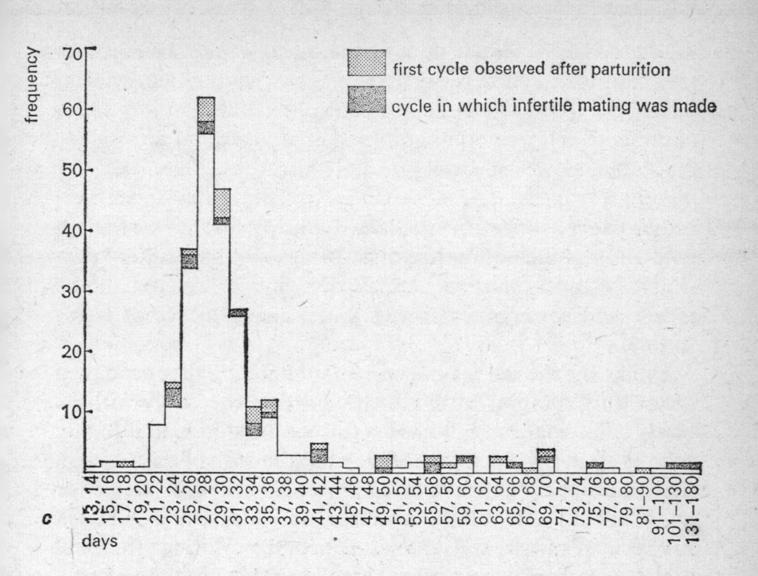

Figure 13 Data on the menstrual cycle lengths for several species show similar skewed-normal distributions

a Women

b Chacma baboons

c Study of *cynomolgus* macaques

in redness of the sexual skin, and about this time she becomes pregnant for the first time. Some females show some swelling during the first cycle or so after each lactation, but in general adult females have only slightly oedematous sexual skin. Other macaques, however, swell as adults also (e.g., *M. silenus*, *M. maurus*, *M. sylvana*, *M. nemestrina*) but all the macaques have rather vague, shapeless swellings, typically centred on the tail root and not including the vulva. Males also have sexual skin which becomes hairless at maturity, and bright red in the rhesus and some other species under the influence of testosterone.

Unlike the rhesus, the baboon infant has a bright pink naked area on its rump at birth which contrasts cheerfully with its black natal coat. In both sexes this skin darkens to a highly reflectant silvery black in the first year (the flash of this patch in the sun probably serves as an important 'here-I-am' signal for all baboons). This skin, however, is not 'sexual skin' in Zuckerman's sense, and it remains unchanged during menstrual cycles. At puberty the perineal area of females begins to swell. The swelling increases during the follicular phase of the cycle and decreases suddenly a day or two after ovulation, and is completely un-oedematous during the last week of the cycle. An adolescent female's perineum may look rather puffy for a month or two before her first swelling, but she may conceive during her first cycle with a full swelling; thus there is no period of adolescent sterility as in the rhesus monkey. This pattern of swelling is found in baboons, mangabeys, and chimpanzees, and is different from that of macaques in the area involved and the timing relationship with the menstrual cycle. The size and shape and colour of swellings are individual characteristics. The size and colour can also vary quite rapidly under the influence of social factors. Thus if a female baboon is attacked her swelling shrinks and becomes paler. When a female talapoin is placed with a male her swelling reddens. Female baboons kept in isolation grow much larger, paler swellings than when they are living in a group (Rowell, 1970a).

In the species which I have handled, the most dramatic colour changes occur during pregnancy. In rhesus monkeys the red

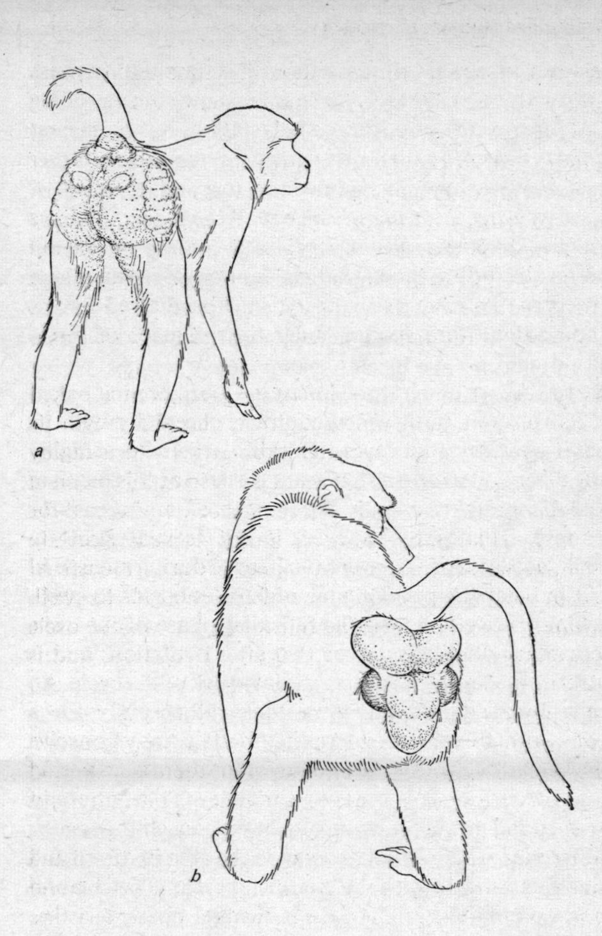

Figure 14 Sexual swellings involve different areas of tissue in different species
a Young rhesus monkey
b Baboon

sexual skin becomes almost incandescent by the end of pregnancy, and redness also becomes more intense on the face, nipples and belly; in baboons the black pigment in the lateral naked rump skin disappears, and this area also becomes brilliant red. The vulval area of mangabeys and vervets and the rump of talapoins also redden during pregnancy. In baboons, mangabeys and talapoins the skin which swells during menstrual cycles becomes very much reduced during pregnancy so that it looks quite different from the non-swollen phase of the cycle. In all cases the bright red fades immediately after birth to a pale pink and in baboons the black pigment slowly returns during lactation. From a distance, the rump of a pregnant female of all these species is much more noticeable than that of a female in oestrus, but no one has yet suggested why the fact of pregnancy should be so well advertised, although it is generally assumed that bright colours do have a signal function.

Most Old World monkeys have a gestation lasting five to six months (compared with around two months for carnivores of similar size), while the gestation of New World monkeys (*Cebidae*) and apes is longer, about seven to nine months where it has been measured.

After birth the reproductive tract becomes very small and thin walled and does not undergo menstrual cycles. The female is said to enter a 'lactation interval', which is a rather misleading term because it is not usually terminated at weaning – indeed most monkeys do not wean their offspring completely but nurse them through the following pregnancy for decreasing amounts of time. In baboons and some other species, the perineum begins to swell again after two or three weeks, should the infant die, and normal cycling is resumed at once. If the infant survives, the lactation interval appears to vary with locality (due either to nutrition or through genetic differences related to local food supplies): five months in Uganda, eight or nine in the Gombe Stream area, over a year in Nairobi. The differences may be due directly to differences in nutrition, or be genetic differences indirectly related to local food supplies. Social factors may also control lactation interval: in caged groups high-ranking females tend on the whole to begin cycling sooner than low-ranking ones, and to get pregnant earlier (this is a casual observation that has not so far been proved).

The purpose of this survey of anatomy and physiology has been to show the diversity of reproductive processes among monkeys and to place the menstrual cycle into its context of the whole female reproductive cycle. We shall now return to consider changes in the relationship of individual pairs of monkeys with the female's menstrual cycle.

Mating behaviour

What we are really discussing here is the concept of oestrus. This should be a strictly behavioural term signifying willingness to mate, exactly equivalent to the vernacular 'heat'. During the earlier phases of the development of reproductive physiology, willingness to mate seemed to be closely correlated with ovulation (as it is in many domestic animals) and the term oestrus acquired a physiological connotation such that in 1932 Zuckerman had to remind his readers that it *also* had behavioural implications! More recently it has become clear that behavioural and physiological phenomena are not related in precisely the same way in all species, and oestrus and ovulation are again being considered separately.

Thinking about oestrus in primates has also gone through changing fashions. In 1900, Heape pointed out that primates were willing to mate for much more of the time than were familiar domestic animals. By 1931 Miller was claiming that primates were always willing to mate, though admitting that some species were even more willing to mate at some times, as for example baboons when they were swollen. He suggested that continuous availability of females for copulation was the basis of primate social grouping and was a basic feature of primates, related to their unique menstrual cycle. Then Zuckerman and Parkes (1932) described the menstrual cycles of baboons, relating endocrine changes with swelling and the rather clearcut oestrus period of these animals. The rhesus monkey was also becoming a laboratory animal at this time, and though early reports described the females as being receptive throughout their cycle (Hartman, 1932), fluctuations in receptivity were reported shortly afterwards (Ball and Hartman, 1935). It then became fashionable to assume that all primates except man show limited oestrous periods around ovulation under normal circumstances, although in the boredom of captivity copulation might be observed at other times. I use the word 'fashion' advisedly, because there were no controlled comparative studies on which to base the assumption, and such studies have only recently been attempted.

A thorough quantitative study of the relation between hormones and intersexual behaviour in rhesus monkeys was initiated by Michael and Herbert (1963) and has since been continued by several workers. A review of the progress in the area to date may be found in Herbert (1970) and separate references will not be given here. These studies have been carried out under carefully controlled conditions, interactions between pairs of variously prepared animals being observed for standard test periods of an hour or less, the monkeys being caged individually for the rest of the time.

Michael and Herbert first reported that grooming between male and female rhesus monkeys in their test situation varied with the menstrual cycle of the female: In the middle of the cycle the male groomed the female for the most part, while at the beginning and end the female groomed the male. The change in male's grooming was paralleled by a midcycle increase in the number of ejaculations per test, and other measures of sexual activity. Thereafter work was continued with ovariectomized females. There was little sexual activity in pairs where the female was ovariectomized and untreated. Oestrogen therapy for the females increased sexual activity of the males to midcycle levels, while treatment with progesterone in addition to oestrogen reduced it again. Treatment with progesterone alone did not increase sexual activity either. Thus external sources of hormones could be used to duplicate changes observed in the normal cycle. While the sexual behaviour of a male increased when his ovariectomized partner was given oestrogen, her behaviour changed little. It was therefore possible to separate two factors, the females' attractiveness and their receptivity. Only their attractiveness was dependent on the female sex hormones. Oestrogen rubbed on the sexual skin, which pales after ovariectomy, restored its bright colour but had no effect on the male's response, so that the colour was not the relevant signal. On the other hand males were much attracted to untreated ovariectomized females on whose rumps the vaginal secretions of an oestrogen-treated female had been rubbed. Thus attractiveness seemed to be based on an olfactory stimulus, a smell produced by the vagina under the influence of oestrogen. (This is a pheromone.) Receptivity of the female was found to depend on testosterones secreted by the adrenal cortex.

Herbert extended the testing procedure to include a second female in the test cage. In these circumstances the male would show a marked preference for one female, and this pair would groom and mate together, while the second female sat in a corner and was occasionally threatened by the consort pair. By manipulating known individual preferences and the dosage levels of hormones the females were given, Herbert and his co-workers were able to show that 'the hormones of the adult are only one factor regulating social activity. Social factors are also important, and in these observations the determinants of male preference could not be understood.' In other words, as soon as there are more than two animals in the situation, it is no longer possible to predict the pattern of interaction from a knowledge of the hormonal state of the individuals involved alone, or not with any great accuracy.

Of course the short test situation with three animals is still a long way from the continuous interaction of many individuals which occurs in a normal free living rhesus group. In the wild it is extremely difficult to obtain sufficiently accurate data on reproductive cycles to compare with the laboratory findings, though Saayman (1970) has been able to make reliable observations, and even to perform experiments on the sexual behaviour of a troop of chacma baboons in the Transvaal. Sexual behaviour in rhesus monkeys is being extensively studied in the free-ranging rhesus colony on Cayo Santiago. The original report on this colony shortly after it was established (Carpenter, 1942) described a midcycle oestrus averaging 9·2 days in length, with sexual interaction practically confined to these periods. In fact the 45 oestrous periods observed by Carpenter were evenly distributed in length between 4 and 15 days, and, more important, their midcycle timing was inferred; menstruation was not recorded. Recently Loy (1970b) found, in the same colony, that perimenstrual oestrous periods were as common as midcycle ones, and did not differ from them in length nor in the intensity of sexual behaviour. Conaway and Koford (1965) recorded sexual behaviour of the females of one of the Cayo Santiago troops throughout an entire breeding season. They found periods of oestrus alternating with anoestrus, but the timing was so varied that it is difficult, looking at their data, to relate the behavioural cycles with possible hormone fluctuations during menstrual cycles; some older females were almost continually in oestrus throughout the whole season, while some young animals showed only a few brief receptive periods. Furthermore, after the infants had been born the authors counted back and found that nearly half of their observed oestrous cycles had in fact occurred during pregnancy – this was especially true of the older, high-ranking females (with whom, incidentally, the highest ranking males mainly consorted, thus 'wasting' potential opportunities for procreation with other females). Loy also reported that in his sample all females showed several oestrous periods after conception. This data is very difficult to relate to the ordered effects of hormones found in the laboratory studies. In particular the regular occurrence of oestrus and consort behaviour during pregnancy is hard to reconcile with the laboratory finding that small amounts of progesterone suppress female attractiveness. It seems that the social regulators of sexual activity demonstrated by Herbert and co-workers may become still more important, and hormonal factors relatively less important, in normally complex monkey groups.

Some of the non-hormonal determinants of sexual behaviour of rhesus monkeys were investigated by Kaufman (1965) using the same Cayo Santiago troop that had been observed by Conaway and Koford. He was not only considering mating, but the complex of behaviour seen in oestrus, including following and grooming, approaches and 'friendly' gestures, between males and females, and hyperactivity in the female. Most of these behaviour patterns are not unique to oestrus, but are much more frequent then. Usually a female will interact intensively in this way with one male at a time, forming a 'consort pair'.

Kaufman found that age was an important factor in determining the amount of time that females were receptive in a breeding season: Up to seven years, the older the female the longer her receptive periods and the more of them she showed. The first females to start breeding each year were parous lactating females with no surviving infant from the previous year, and the last to start were the adolescent females entering their first breeding season. The number of males with which the female consorted during each oestrus also increased with age up to seven. The attractiveness of the females, as indicated by the amount of following and grooming by males, increased rather earlier than receptivity, reaching a peak at about five. Age did not have any effect on sexual activity of males.

In Kaufman's study there was a correlation between sexual activity and rank in males, the higher ranking males being most active, consorting for longer with more females. He could not find any evidence that high-ranking males were more active at any particular stage of the females' oestrus. (There is a persistent suggestion that high-ranking males take over oestrous females at the time of ovulation and thus leave most offspring. This point will be returned to in chapter 9.) Conaway and Koford did not find a correlation with rank and sexual activity in the same troop but some of the males had moved between troops since their observations.

In contrast to males, females' sexual activity was not related to rank. Kaufman also found differences in level of sexual activity between individuals of both sexes that were not related to age or rank, and which for the moment must be ascribed to 'individual differences'. Most females were consistent in their behaviour from year to year, apart from the changes with age already discussed, but some changed, for no known reason.

As well as general levels of sexual activity, any selection of partners based on individual recognition rather than general characteristics will also be biologically important. Kaufman did not detect any persistent partnerships in his study that could not be explained by the general trends he found. There seems, however, to be an 'incest taboo', an inhibition of mating between sons and mothers. Sade (1968) pointed out that son–mother matings were much less frequent than would be expected with chance association in Cayo Santiago, in spite of the fact that sons who lived in the same group as their mothers interacted with them rather frequently. Son–mother mating was also not seen in Japanese macaque groups whose genealogy is known.

For the healthy rhesus monkey then, we can make a list of the determinants of sexual activity so far described: hormonal levels, social companions, age (in females), rank (in males), breeding success (in females), general environment (determining breeding season or its absence) and 'individual differences', which presumably includes a number of genetic and experiential effects.

Nowhere near so much data is available for any other species, and comparable studies are only just being attempted. It is already possible to say, however, that the rhesus cannot be taken as typical of all monkeys. We shall consider the comparative data available for each of the determinants listed for the rhesus monkey.

Hormonal levels

Clearcut relationships between mating and the menstrual cycle are seen in baboons and mangabeys, and the distinction between oestrus and non-oestrus is sharper than in the rhesus. Mating is almost exclusively with swollen females, and is mainly confined to the last week or two of the swelling when it is at maximum size. Sexual behaviour in Ugandan baboons ceased abruptly the day before any decrease in swelling size was apparent, but in chacma baboons (Saayman, 1970) activity decreases gradually with the swelling. Chimpanzees have a sexual swelling with roughly the same relationship to the menstrual cycle as that of baboons, but oestrus is less sharply delimited in this species, and copulation with slightly swollen or non-swollen females seems to occur more often than in the monkeys. Unlike the baboons, swellings, with oestrous behaviour, occurred in pregnant chimpanzees (van Lawick-Goodall, 1968). In vervet monkeys no relationship was found between female reproductive state and sexual behaviour; apart from a few weeks after the birth of an infant, most females appeared to be in permanent oestrus. Sykes' monkeys showed alternating periods of oestrus and anoestrus, but these were not consistently related to menstrual cycles – each female tended to be most receptive at a different stage of the cycle. There was some mating during pregnancy in this species (Rowell, 1970b). Since Sykes' monkeys behave like this, field reports of midcycle oestrus must be treated with caution where menstruation could not be recorded, because the finding is almost always actually only of alternating periods of oestrus and anoestrus. It is occasionally possible to detect menstruation in the field, however, and Jay (personal communication) found that oestrous periods in langurs did occur midway between menstruations when these could be seen.

Effect of male preference on sexual behaviour

There is no comparative data on the effect of male preference on sexual behaviour described by Herbert, although at the anecdotal level such preferences are generally known among captive primates.

Strong pair bonds within a baboon troop have been reported by Ransom (1971). They seem to develop slowly, since no nulliparous or primiparous females were involved in pair bonds, but only multiparous older females. In some cases the mating preference of the pair was followed by an equally intense 'paternal' relationship between the male and the females' subsequent offspring.

'Incest taboos' as described for macaques are not known from other genera of monkeys because there are no available groups with known genealogies. Van Lawick-Goodall has not seen son–mother mating in the Gombe Stream chimpanzee population, in so far as relationships between adults are known there, so it is likely that a similar inhibition pertains.

Age

Van Lawick-Goodall (1968) found that older chimpanzees were both more receptive and more attractive than younger ones, and old female baboons had longer oestrous periods than younger ones in my caged group. On the other hand old Japanese macaques were less persistently sexually active than younger ones (Hanby, Robertson, and Phoenix, 1971) (there may be a difference in the criteria for 'old' here). In hamadryas baboons, Kummer (1968) found that intense sexual activity was characteristic of young mature adult males; both younger and older males were less active, the younger ones apparently because of lack of opportunity in the harem-organized society, the older males because they became increasingly involved in activities related to troop leadership.

Sexual activity and rank

There is a general tendency in caged groups for high-ranking females to interact more with an adult male than low-ranking ones, and to breed more; the lowest ranking female of a captive group is quite often a non-breeder. In view of Kaufman's findings on free-ranging monkeys, it is likely this is a cage artefact, one of the many effects of the stress of captivity. A reverse causal effect has also often been suggested, that females change in rank during menstrual cycles, rising with oestrus and falling as oestrus ends. Several quantitative studies have been unable to substantiate this, and it seems to have arisen from a somewhat androcentric definition of high rank by proximity to an alpha male – an obvious tautology when oestrus is being considered. Some vervets, baboons, and chimpanzees (Young and Orbison, 1944) do become involved in more aggressive interactions during oestrus. Rhesus monkeys, on the other hand, are involved in more fights around menstruation (Rowell, 1963). In males, DeVore (1965) found a general, but not a detailed correlation between sexual activity and rank in baboons. In langurs, Jay (1965) reported that females tended to solicit higher ranking males more at the height of receptivity; her figures indicate that the alpha male copulated most frequently and was second most often solicited, but again amount of sexual activity was not accurately correlated with rank. In captive groups it is common practice to include only one adult male in a group, and in several species which have been observed in the wild, a single adult male moves with a group of females; in these situations the question of the effect of rank on sexual activity is irrelevant. It has been generally assumed that the position of a single adult male in a group is achieved by competition with other males and so is itself an indication of high rank, but with the exception of Sugiyama's (1965) study of langurs, evidence on this point is not available. Among chimpanzees adult males amicably shared copulation, apparently without reference to rank (van Lawick-Goodall, 1968); among ring-tailed lemurs the normal rank order of males in the group was completely changed during the brief mating period, and a normally low-ranking animal won fights and performed the majority of copulations seen; the males reverted to their original ranking when the season ended (Jolly, 1966). In summary, there is general disagreement between studies of different species, and even, as we saw for the rhesus monkey, between different studies of the same species on this point. Probably there is a general correlation, in monkeys, between high rank and high sexual activity, but not a causal relationship, since low-ranking highly active males and high-ranking inactive males are not infrequently reported.

Environmental factors

The effect of weather and other environmental factors in determining overall levels of sexual activity has already been discussed.

Individual differences

Individual differences in sexual activity have generally been commented on, but again there is a lack of quantitative information. Saayman (in press) in an experiment in which he captured wild adult females and reintroduced them into a group after ovariectomy and oestrogen replacement implants observed individual differences in amount of sexual behaviour which were not related to the known dosage of oestrogens each was receiving, above a certain low threshold.

Studies of the control of mating behaviour in primates are in an exciting phase, and this rather inconclusive account of current knowledge can only give an indication of the fascinating complexity which is being revealed as old assumptions are being challenged and more comparative material becomes available; the complex interrelationship of endocrine, social, and environmental factors could become a model of the control of animal behaviour.

We began this chapter with the idea that sexual behaviour was the cohesive factor in the development of complex permanent societies of primates. Recently it has been fashionable to dismiss this old theory, citing the seasonality of breeding in most species, and the relatively brief receptivity of females. Aggression between male macaques and baboons associated with the presence of receptive females has been used as evidence that sexual behaviour is actually disruptive of the social order, and the attractiveness of infants to all other members of the group has been suggested as an alternative cohesive element (Washburn and DeVore, 1961). Both these suggestions are clearly oversimplifications: If breeding is seasonal, then the attractive infants will also be available for only limited periods in the year. If the cohesive function of sexual interaction is denied, we are left with the task of explaining why monkeys and apes are receptive proportionately more of the time than are other mammals – even other social mammals. Speculation about the evolution of social behaviour is a delightful pastime, though it should not be taken too seriously, so the following suggestion might be used as a starting point for discussion. The original primate ancestor presumably was a relatively asocial animal with limited oestrous periods, as are many modern insectivores (the most closely related group of mammals). An extension of female receptivity may have allowed the initial development of longer-lasting associations between males and females. Once a group-living life-style has developed, evolution of female receptivity patterns seemed to have diverged. In some cases, the increase in receptive time seems to have continued, so that as in man and the vervet monkey, mating is almost always possible. Copulation in these species is relatively infrequent, however, when the measure is 'mating per hour', and is accompanied by very little change in other friendly behaviour. In other cases, receptive periods are relatively infrequent, but during them copulation occurs very often, and the whole pattern of interaction is altered as the female forms consort relationships. In some species with this pattern, of which the baboons are the best known example, perineal swelling during the receptive period makes it even more obvious. In the former case it is reasonable to assume that sexual activity is cohesive, in the latter it seems disruptive; indeed it would seem that mechanisms for maintaining group stability would have to be developed to a high degree of sophistication before such potentially disruptive behaviour could evolve without destroying the social structure. Comparison of baboons and guenons (Rowell, 1972) suggests that the former do indeed have a more highly developed repertoire of conciliatory gestures, allowing them to resolve social tensions more successfully, at least in captivity.

7 Reproductive Behaviour: Infant Social Development

Like all mammals, the infant monkey is totally dependent on its mother for some time after it is born, and it is essential for its survival that certain interaction patterns are immediately established between them. Mother monkeys carry, feed, and clean their newborn infants, and defend them if necessary, and the infant must cling and nurse, and indicate any discomfort by movements and noises. In these processes, which lead to the infant's healthy physical growth, it is quite possible for a human caretaker, or even a carefully prepared nursery cage, to replace the functions of the mother. But a monkey also peers intently into her new infant's face, touches its face with her lips, cuddles it and croons to it, and with these activities she starts the learning processes through which the infant develops into a psychologically normal adult.

Monkeys are adapted to live socially just as they are adapted to live in trees, or to eat a certain type of diet. For all these adaptations, in the life of the individual monkey, some aspects are genetically programmed, so that each grows the right sort of limbs and teeth and makes the right communicative gestures; and other aspects the individual learns as it develops, like how to run on lianas, how to recognize food, or how and with whom to communicate. There is no hard line that can usefully be drawn between learned components of behaviour on the one hand and innate ones on the other: some things are much easier for a monkey to learn than others, showing that there are innate 'guidelines' for which responses will be acquired. For example, monkeys learn to recognize other individual monkeys more quickly than they learn to distinguish geometrical shapes.

Isolation and model mothers

Harlow and co-workers at Wisconsin developed a technique whereby newborn rhesus monkeys could be taken from their mothers and reared with minimum amounts of handling in isolation. (This whole breeding and rearing programme was a triumph of animal care techniques which has perhaps not been sufficiently appreciated.) The original object of this programme was to be able to control early experience in the study of the development of learning processes, but as so often happens, a side effect of the original research project became more famous than the main line of thought. For the sake of hygiene, the original infants were kept in wire-mesh cages, but they cried continually (using the long drawn-out call of an infant separated from its mother). Crying was reduced if the infants were given soft cloths, to which they clung tightly. Harlow showed that this clinging formed a basis for an 'affectional' relationship with a model mother – infants allowed to cling to a cloth-covered model were later able to derive reassurance from the model in a stressful situation, while those reared on wire 'mothers' without the cloth covering were not reassured by their model's presence in the same situation (*see* Figure 15). The infants were fed from nipples protruding through the models, but the reinforcement provided by feeding did not affect the affectional bond: Infants given a wire and a cloth model, but fed only from the wire model, still ran to the cloth model for comfort – indeed in their home cage they went to the wire 'mother' only for milk, and clung to the cloth 'mother' for the rest of the time (Harlow and Harlow, 1965, review). An infant baboon which I reared showed an even more complex differentiation of stimuli which would normally all come from the mother. Like the infant rhesus monkeys, the baboon clung to its 'cloth mother' (myself, or the apron in which he was carried, as a second choice) for reassurance. The baboon was provided with a dummy teat which he frequently sucked, and this was also sought in a stressful situation. The bottle from which he sucked milk, on the other hand, provided no reassurance. Offered bottle and dummy he chose the bottle only when he was hungry, and rejected it in favour of the dummy as soon as he was satiated. Thus the satisfaction of sucking was quite clearly differentiated from that of obtaining food and satisfying hunger (Rowell, 1965).

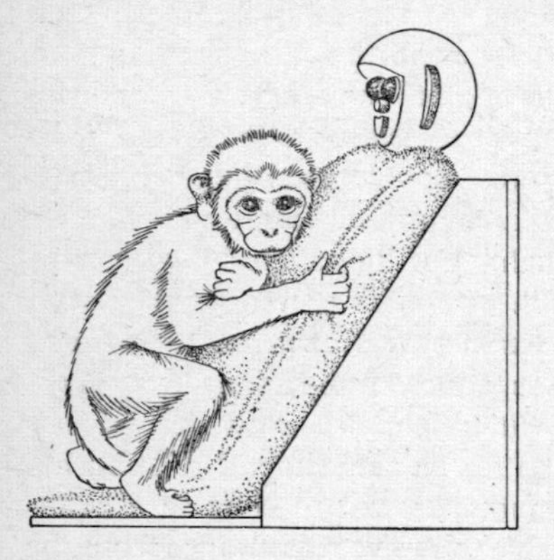

Figure 15 Cloth-covered, wire-mesh monkey, such as is used for rearing young

In the early stages, Harlow and his associates seemed to have invented a 'mother' for infant rhesus monkeys that was better than a real one. The infants grew well, and unlike the infants reared with their own mothers they were never rejected, threatened, or treated roughly by their cloth model mothers. (They cried a lot still, and sucked their fingers, toes, or penis, but the significance of these habits was not at first realized.) When these isolated infants were placed with other monkeys, however, it became obvious that they were by no means normal. They were either extremely aggressive, or they withdrew as far as possible and tried to avoid all contact with another monkey. The normal patterns of interaction between juveniles, play, grooming, and some mild threat exchanges, were completely missing. Later they showed no correct mating behaviour; the males never overcame this inability, but some of the females were eventually impregnated, and then, when their infants were born, they showed no appropriate maternal behaviour. Instead of cuddling their infants they pushed them away, seemed frightened or at best unaware of them. (In later pregnancies these females improved their maternal performance to some extent.) In short, the isolated infants, in spite of their good physical development and normal problem-solving abilities grew up to be socially totally incompetent.

Taking far less elegant evidence from a quite different source, we can suggest that some sort of experience with conspecific is generally required by infant monkeys and apes if they are to show normal reproductive behaviour as adults. Many primates are captured as infants and hand-reared as pets, then transferred to zoos when they get older. Attempts to use such animals as breeding stock have a very low rate of success: typically these ex-pets show little or no mating behaviour, and the females do not care for their infants, which are then hand-reared, thus creating the breeding problems of the next generation. The histories of these zoo animals are of course very varied, and the fact that some of them do breed suggests that one might be dealing with quite short critical periods during which social experience is necessary. This was investigated in the rhesus infants. Isolation for the first three months produced no permanent effects if the infants were then placed in a group, but isolation for the whole of the first year destroyed all social ability. The period between three and nine months seemed to be critical for establishing normal social behaviour, although the communicative gestures have nearly all appeared in the first three months.

Total isolation is an extreme procedure that predictably produced extreme effects. Attempts to analyse the social learning processes whose existence was thus established have been of two main approaches: Harlow and his school have been gradually adding components of a normal environment to the complete deprivation situation, while Hinde and others started from a more or less complete social situation and investigated the effects of subtracting single factors. Both these approaches have provided important insights into the social determinants of infant development, though they are complicated by secondary effects. For example adding or removing an adult female to the infant's social environment not only affects the infant's own interaction with that female, but she also interacts with and modifies the behaviour of the infant's other companions: in other words, the social environment is more than the sum of its components. None the less for the purposes of explanation we must consider the parts separately, while bearing in mind the woven fabric with which we are really dealing.

Mother–infant

The infant's first and for a long time its major social partner is its mother – and we have seen already that this relationship lasts long beyond the period in which it is physiologically necessary. Because they stay together, and both help to maintain proximity, the infant takes its mother as its first model for imitation.

Thus it learns her avoidance habits, both of potentially harmful objects like snakes and of higher-ranking individuals, and her food habits. Van Lawick-Goodall (1968) has described how young chimpanzees learn how to use tools by watching their mothers. Infants also imitate other members of the group – squirrel monkeys seem to learn mainly from older juveniles rather than their mothers (Baldwin, 1969), while in Japanese macaques the process may be reversed, as mothers learn new food sources from their more inquisitive young (Itani, 1958). The pattern of mother–infant interaction changes, of course, with time, as the infant grows and becomes more mobile. One of the first jobs has been to describe, first qualitatively and then quantitatively, this changing relationship. Figures 16 and 17 are examples of quantitative descriptions, and they may themselves suggest points which will repay further study; for example Figure 17 suggests that an important group of changes might occur around eight weeks in baboon development, where the graphs cross. Infant development has frequently been described in terms of 'stages' (e.g. Hansen, 1966; Jay, 1963). This can be convenient in field work, but for analysis it seems more useful to stress the continuity of the developmental process, since each part has a different time course, and each may be differentially influenced by the environment.

Effect of environment

Comparison of these quantitative descriptions between individuals of the same species and between groups of different species have shown some ways in which the environment may affect the interaction pattern of mother and infant. Comparison of four species of African monkeys held in similar cages showed that infants of arboreal species were much slower to leave their mothers than infants whose mothers spend much time on the ground, and this difference cut right across taxonomic divisions. Vervet and baboon infants would leave their mothers and walk on the floor in their first week while Sykes' monkeys and white-cheeked mangabeys stayed in contact until they were over a month old, and when they did leave they climbed first vertically. Restriction by the mothers was not the controlling factor (Chalmers, 1972). There would seem to be an obvious advantage in infants remaining on the mother and avoiding the risk of falling until they are better coordinated.

Pigtail macaque mothers and infants interact differently in 'rich' and 'poor' environments. The difference was in the opportunity to play with a variety of toys which the rich environment provided for the infants, while the poor environment was a bare cage, in a sound-proofed room. Rich environment infants spent more time away from their mothers and oriented more behaviour towards the environment and less towards their mothers and themselves. Mothers in a rich environment were less aggressive towards their infants, perhaps because they were not themselves the sole available play object, and rich environment infants responded to maternal aggression by leaving her, while in the poor environment infants responded by clinging quietly to the mother. (These differences were found at one extreme of environmental poverty, and infants in the poor environment were also physically retarded. A cubic metre of space, and only the mother as companion, might be accounted fairly impoverished for a macaque even if toys were available.) (Jensen, Bobbitt and Gordon, 1968.)

In baboons, one of the more striking changes in mother–infant interaction is the move from being carried always

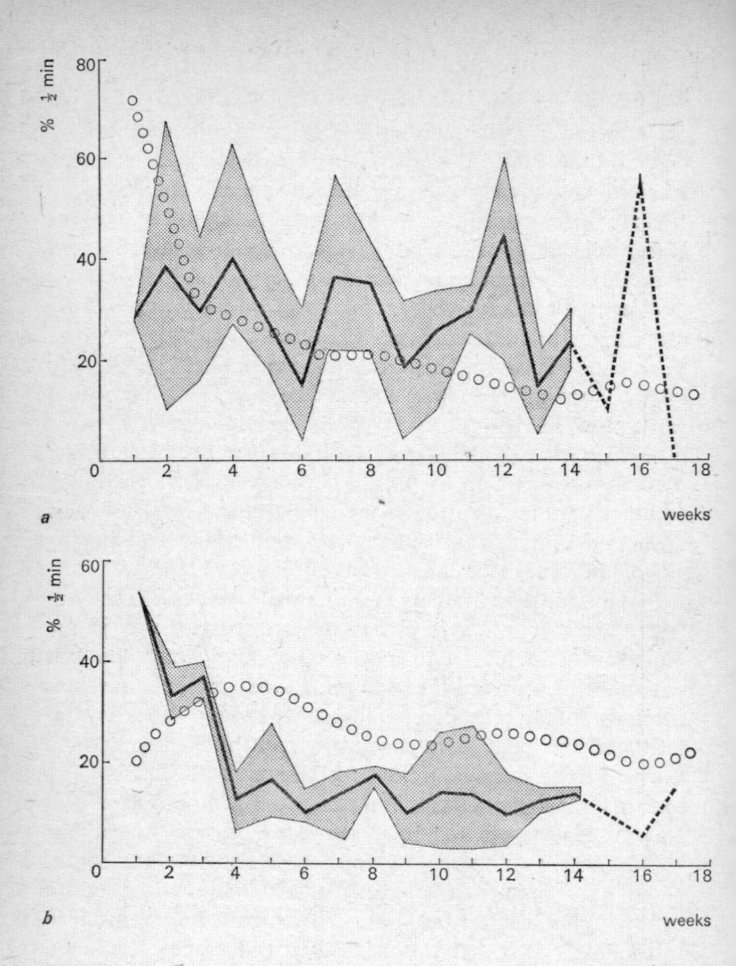

Figure 16 An example of a quantitative description of mother–infant interaction which demonstrates the rate of development. The dark line shows the mean, the shading the range of values for four infant baboons. The use of rings is the mean score for six rhesus infants
a Time asleep
b On nipple, eyes open
c Within arm's length
d Out of reach of mother
Source Rowell, Din and Omar, 1968

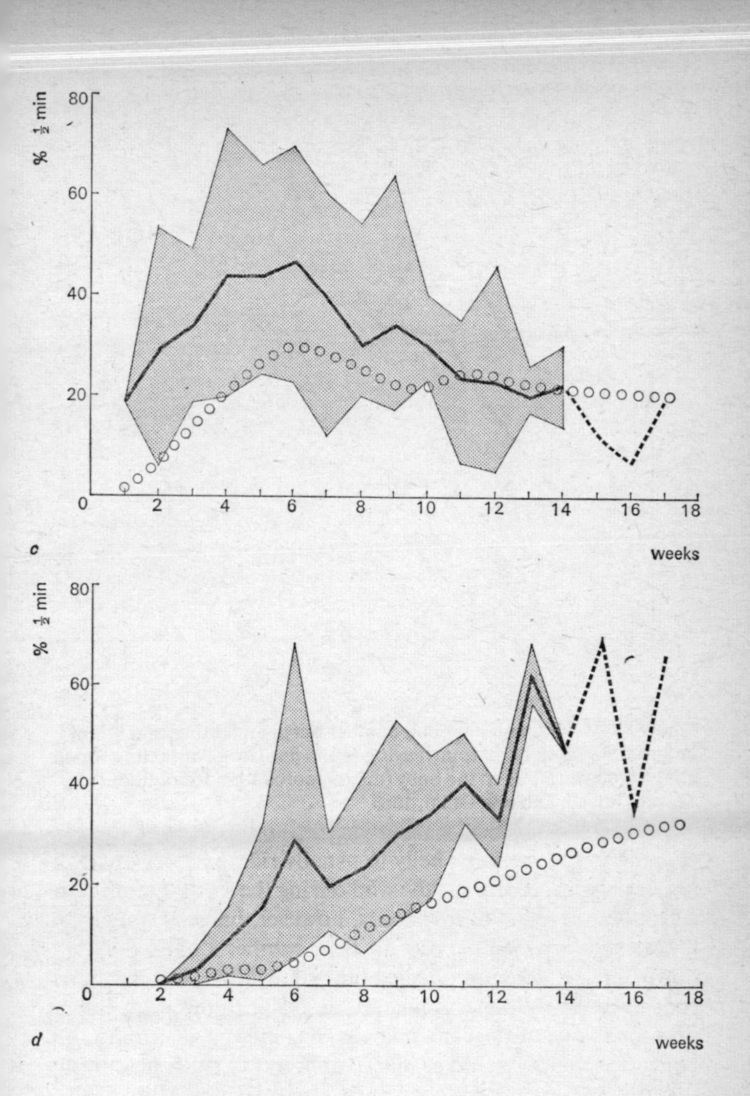

% ½ min
80
60
40
20
0
2
4
6
8
10
12
14
16
18
c
weeks
% ½ min
80
60
40
20
0
2
4
6
8
10
12
14
16
18
d
weeks

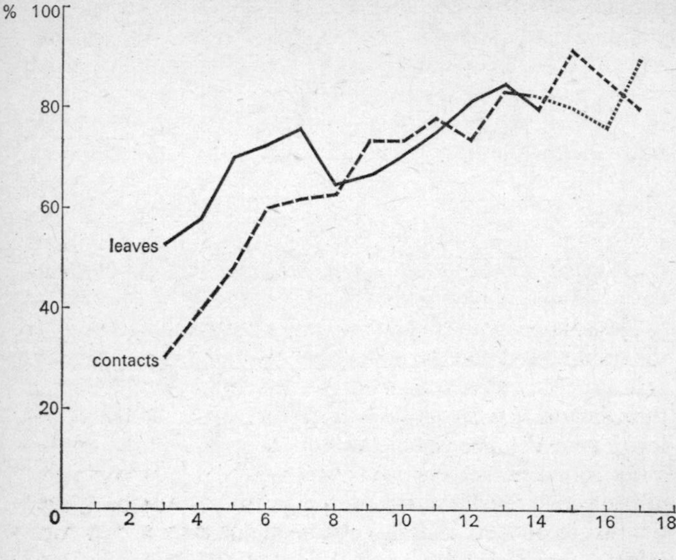

Figure 17 Making and breaking contact between mother and infant baboon. Percentage of total leaving (solid line) and contacting (open line) which were due to the baby (mean scores from four infants)
Source **Rowell, Din and Omar, 1968**

clinging to the mother's belly to usually riding jockey-style on her back, which occurs in the wild during the second month. In a cage with a series of low doors, however, babies rarely rode jockey-style – when carried on their mother's back they lay down, and so were not brushed off by the lintels, but they also continued much longer riding on her belly. In the wild, riding position is one of the early sources of conflict between mother and infant, which would be missed entirely in a study made only in captivity.

My Sykes' monkey cage was made of mesh large enough for small infants to creep out, and by six months of age they returned to their mothers only briefly during the day, spending the rest of the time in nearby trees. At last they had to be confined because they were about to become a nuisance; when they could no longer leave the cage (which was large, and contained nine older monkeys) they reverted to clinging and suckling like babies several months younger. It is clearly important to know the parameters of the physical environment in which mother–infant interactions take place.

The social environment

Hinde and Spencer-Booth (1967) compared the behaviour of isolated mother–infant pairs with that of group-living pairs.

The isolated mothers lived in identical cages to those housing the groups, and could see and hear but not touch the group animals. The pattern of interaction of isolated mothers and their infants was significantly different from that of group-living animals. The isolated infants spent more time further away from their mothers earlier; this seemed to be due to a lack of protective restrictiveness on the part of the mother. Later (in the second six months) isolated mothers rejected their infants' approaches more often; the isolated infants pestered their mothers more because they had no alternative playmates. Restrictiveness in the first six months was greatest in two group-living mothers whose group included an adolescent female who persistently stole their infants, and from whom they could not retrieve them easily because the adult male preferred her and supported her in fights.

In the matter of restrictiveness it is possible that the isolated mothers behaved more nearly like free-living mothers than did those in the group situation. Wild baboon mothers at Ishasha were not seen to restrict their babies beyond an occasional brief instance (typically they use the tail for a leash), but all except the highest ranking, caged female baboon did so, and some so much that in the third month they might spend an entire observation hour fighting their infant as it struggled to be free. If such females were separated from the rest of the group for a short period they immediately relaxed and allowed their infants to play, retrieving them only if they strayed towards the wire separating them from other baboons. It was possible, by separating the mothers and infants with one other member of the group at a time to show which individuals were causing the mother to be restrictive (Rowell, 1968). It was thus clear that the social, as well as the physical environment, has important effects on the mother–infant interaction.

Even in the same environment, mother–infant pairs behave differently from each other. Hinde and Spencer-Booth (1968) were able to show that differences between pairs were consistent over long periods. Differences were due to both mothers and infants, but maternal differences were most important in the first five months, while differences between the infants became more important thereafter. Some of the causes of differences which have been investigated are rank and parity of mother and sex of infant.

Rank and parity are both related to age in long-established groups, so the effects of the two are confused in group situations. The immediate effects of rank are seen in the amount of maternal restrictiveness – high-ranking mothers are more relaxed. DeVore (1963) said that wild subordinate female baboons were more short tempered and less responsive to their infants than dominant females. These are qualitative observations, there has been no study specifically concerned to relate rank with mother–infant interaction directly, though observations on the relationships of both mother and infant to other members of the group (q.v.) have frequently related to maternal rank. The responses of other monkeys to the mother apparently teach the infant a set of expectations of the behaviour of others, and are of course related to her rank. Thus there is evidence that maternal rank determines the future rank of her offspring. In Japanese macaques, females take the rank position immediately beneath that of their mother, and male infants of high-ranking females retain access to the central part of the group instead of moving to the periphery like sons of lower-ranking females. Here they are in a good position to identify with high-ranking males and eventually enter the central hierarchy themselves (Kawai, 1958). However, among the Cayo Santiago rhesus the high status of sons of high-ranking females seems to depend directly on the mother's willingness to intervene on his behalf: when males leave their natal group they do not retain their old rank in their new group.

Mitchell and Stevens (1969) found differences in the behaviour of primiparous and multiparous mothers in a rather stressful test situation: primipares were more anxious and stroked their infants more. He and co-workers also found differences between first born and later born infants at $2\frac{1}{2}$ years: First born juveniles showed more stereotyped movements and other disturbances, and less hostility, play, and self-manipulation. The apparently retarding effect of being a first born was more pronounced in males than in females, and in general, the largest differences were associated with parity in the first three months and sex of infant in the second three months. Primipares differed from old females in a wild baboon group studied by Ransom (1971). In the first few days young mothers seemed to find their infants very uncomfortable – they continually adjusted the infants and would put them down on the ground and move a short distance away from them. Older females were much more relaxed and seemed to have no difficulty in keeping their infants quiet and happy. Ransom felt the difference might in part be due to the difference between the small nipples of the new mother and the greatly extended ones of the multiparous animal. The first-born infant has difficulty in maintaining his hold on the nipple whereas the later born can lie easily back on the mother's lap with the nipple still in its mouth. The nipples also appeared to be more sensitive in the first lactation. The result of this difference in the first few days was that infants of older mothers stayed in closer contact with their mothers for the first few months – they were rarely rejected, but cradled as much as they wanted, while the first-born infants were encouraged to very early locomotor independence and experienced early rejection. At about six months, however, when the first-born infants had established an apparently satisfactory, if rather independent, relationship with their mothers, the older mothers began rather suddenly to wean their offspring and make them walk rather than being carried. The dependent infants found this change highly traumatic, to judge by their frequent tantrums. Thus infants of primiparous and multiparous mothers, though perhaps undergoing equivalent overall amounts of stress, had very different timing in the changes of their relationship with their mother.

Both mothers and other members of the group pay particular attention to the genitals of new-born infants, and inspect them carefully: Baboon mothers investigate the penis of male infants, while macaques handle the scrotum – the most obvious organ in the new born in each case. Rhesus mothers behave differently towards male and female infants even in the first week (Mitchell, 1968). In the second three months the differences change – female infants were restrained more in the first three months; in the second, the biggest difference was that males were punished more than females, probably because the infants' own behaviour diverges as play becomes a more important part of their activity. Male infants initiate more play, and bite and tussle in play more than do females. Goy (1968) injected androgens into pregnant females and found that the play behaviour of genetic female infants was masculinized. Since the the infant's own sex hormone production is extremely low in the first year it seems reasonable to attribute sex differences to the organizing effects of androgens in prenatal development. The experimental females also had modified external genitalia as a result of the androgen treatment however, and perhaps insufficient attention has been paid to possible changes in response by companions as a result of this change in appearance, since we have seen that adults establish a pattern of interaction with the infant after inspection of its genitalia.

The mother–infant interaction pattern can also be modified experimentally by separating them for a short time. Interest in this procedure derives from human experience; children separated from their mothers for short periods (for example when one or the other has to go to hospital) often show profound short term changes in behaviour, and sometimes show serious long-term effects. Attempts have been made to duplicate these effects in monkeys so that they can be studied, as of course experiments on children are impossible. Macaque mothers and infants have been separated for a few days (Seay, Hansen and Harlow, 1962; Spencer-Booth and Hinde, 1967), for a month

(Kaufman and Rosenblum, 1967), and repeatedly over the first 8 months (Mitchell, Harlow, Griffin and Møller, 1967). Timing of the tests, and the living conditions of the animals were also different in the various experiments.

When their mothers are removed, infants first show acute distress, search, make 'lost baby' calls, and scream. Then they become rather inactive, sit hunched, alone, and play little; this phase was described as anaclitic depression like that of separated children, by Kaufman and Rosenblum. In the long separations these symptoms diminished and the infants' interactions with peers slowly returned towards normal. Though the depression was the usual response in rhesus and pigtail macaques, it did not appear at all in separated bonnet macaques. Bonnets are much less aggressive monkeys that habitually spend most of their time huddled together: When their mothers were removed the infants were cuddled and protected so thoroughly by the rest of the group that they appeared to recover quickly from the separation.

When their mothers were returned, the infants immediately clung to them – no monkey infant has shown the rejection of the mother on return which causes so much heartbreak in human families. The mother–infant relationship became much more intense, and in terms of such measures as time spent off mother or at a distance from her, the infants regressed many weeks in development, apparently due mainly to a change in demand by the infants, which repeatedly approached their mothers. There would follow a period of readjustment in which many measures of mother–infant interaction fluctuated violently from day to day. There have been conflicting statements about the long-term effects of such separation in monkeys (as in humans), partly due to differences between individuals and living conditions, and partly due to the sensitivity of the tests employed. Thus Seay and Harlow found the effects to be transient and unimportant, while Spencer-Booth and Hinde found significant differences between separated and non-separated infants up to two years later. Their tests, which were inserted into the normal daily routine of the groups in such a way as to produce minimal general disturbance, showed that the previously separated juveniles were still less quick to recover from alarming situations, and less willing to take titbits from an experimenter. After three years rhesus monkeys which had been repeatedly separated from their mothers in their first eight months were lower in social dominance compared with controls (Sackett, 1968). These effects were small compared with, say the effect of total isolation for the first year, but they are much more comparable to the types of disturbance produced in children.

'Weaning' and maternal rejection

On the whole mother monkeys are remarkably tolerant and permissive of their offspring, and rejecting or non-cooperating behaviour towards their infants is relatively rare. This in itself makes it catch the attention, and more attention has been paid to maternal rejection than its frequency would seem to justify. It is becoming apparent, however, that far from being merely an expression of maternal bile which the unfortunate infant must endure, the appropriate amount of rejection by the mother is essential for normal behavioural development. The first hint of this came from the Wisconsin studies of isolated and mother-reared infants and infants reared together. Clinging behaviour decreased much more slowly in infants reared with cloth models than in those reared by their mothers, and was extremely persistent in infants reared with other infants. When clinging, an infant cannot explore or develop new physical or social skills, and if they cling too much too long they become retarded (Harlow, 1969). The relative importance of the roles of mother and infant in increasing independence was analysed by Hinde (1969). Although to casual inspection it seems as if the infant simply becomes more independent of its mother as it gains motor competence, the mother being a rather passive partner in the process, Hinde was able to show that it is the mother that is primarily responsible for the increasing time the infant spends out of contact with her, and at a distance from her. Even in the first five months, when it was the mothers who were primarily responsible for maintaining proximity, they were also promoting the infants' increasing independence. As Hinde points out, these changes in the mother's behaviour may in turn be the result of changing stimuli she receives from the infant as it grows; it is only possible to separate their roles to a limited extent in a first-stage analysis.

The precise function of maternal aggression itself in the changing relationship has had different interpretations. Hinde sees weaning as leading to increased independence. Some of the cloth mothers on which the Wisconsin rhesus were reared could punish their infants, by emitting an air blast, and the infants responded by clinging more tightly to the model. Kaufman and Rosenblum (1969) also found that punishment by the mothers of pigtail and bonnet infants was associated with increased amounts of clinging. Pigtail mothers punish much more than bonnet mothers, and yet pigtail infants are much more dependent on their mothers. Kaufman has recently found that pigtail social organization is based on strong permanent mother–infant ties leading to the formation of socially exclusive matrilines, while bonnet macaques develop a much less structured society as the colony matures (personal communication).

Siblings

In wild groups of baboons, macaques, vervets and chimpanzees (the only species on which sufficiently long-term studies have been made) the next most frequent partners for social activity for the infant after its mother are its maternal siblings. Even the more fearful mother who prevents contact with her new infant by all other monkeys will allow investigation by her own elder children. In newly formed captive groups a mother may use an unrelated adult female as babysitter, but where the matriline is established this will be the job of a young female relation. Spencer-Booth (1968) found that female siblings two years older than the infant interacted with it more than anyone else other than the mother. The effect of this sibling relationship is most important in the infant's interaction with male juveniles however: whereas young female baboons are attentive to all new infants, young males pay them very little attention. They will however investigate and play with their own younger siblings to some extent (Ransom, 1971), so any experience of young males an infant has will be almost entirely with his brothers. In several macaques the sibling relationship continues to be the main basis of partner selection for grooming and other friendly behaviour into adult life. This is probably more widespread in monkeys than is currently realized, since most field studies are not long enough to recognize genealogies, or even individuals, and a reported general interest by juvenile females in infants may well turn out, on further investigation, to be a series of specific sibling relationships. It is not universal, however: Kaufman has found that bonnet macaque groups do not organize themselves on this basis (*see* chapter 8 also).

Because most studies are short, it has usually been only juvenile siblings which have been identified with any certainty, and sibling relationships have been mainly considered in relation to peer relationships of infants and juveniles.

Peers

After about the first six weeks, young baboons and macaques soon come to spend most of their waking time in play with other infants and juveniles, including siblings. Play begins with exploration of other infants and objects and increases in complexity as the infants gain physical competence. It comes to include such a wide variety of motor patterns that it has so far proved impossible to define objectively, yet it is easy to recognize, and is an important concept in understanding social development. Many of the components of adult sexual and aggressive behaviour first appear in play sequences, and practice in their use seems to help the infant grow into a socially competent adult. Loizos (1967) goes further and suggests that play interaction of young primates may be functionally equivalent to the rapid imprinting process seen in newly hatched nidifugous birds whereby species identification is made. Thus Harlow (1969) found that infant rhesus reared with model mothers but allowed only twenty minutes a day of play with peers grew up to be normally socially competent – better, in fact, than infants reared with only their mother as companion. They did not, however, develop as rapidly as infants reared in a group of adults and other infants. For example, we saw that isolated infants, raised with models, did not show normal sexual behaviour when adult. Infants allowed play experience showed normal sexual behaviour by fifteen months, while group reared infants showed 'correct' mounting and presenting for mounting by fifteen weeks.

Baboons tend to form 'maternity bonds' (Ransom, 1971) with other females which give birth at about the same time. They sit together and groom each other frequently, so that when their infants begin to explore they find each other as first play companions. Mothers and daughters also tend to stay together as adults, so that there is a strong chance that the first playmates of an infant will be its cousins. Choice of playmate was analysed by Fady (1969) in a *cynomolgus* macaque group: Young monkeys tended to play with others of the same age group, with siblings rather than others of comparable age, with others from the same 'maternity club', and also infants of high-ranking mothers tended to play together, though low-ranking infants did not select playmates by mother's rank. Different factors will be more important in species with limited breeding seasons, where division of young into age cohorts is very clear as compared to species or populations with year-round breeding activity, where many infants will not have close peers.

The choice of playmate, often based on partly fortuitous circumstances, is important because relationships developed in play with peers will often continue into adult life – the infant does not practice the adult social behaviour patterns *in vacuo*, but in relation to specific individuals. Thus the rather tense, formal relationship between adult males in a macaque group does not seem to be decided at puberty but grows out of the existing relationships of the juvenile males even though these seem qualitatively different, relaxed and playful.

Aunts

We have used this word to include females and older juveniles living in the same group as mother and infant. It does not imply blood relationship (but is used in the same way as the form of address 'aunty' for women friends of the family which used to be taught to children). The category would overlap with that of siblings in many studies.

In all species females take great interest in a new-born baby, peer at its face, touch its fur, inspect its genitals and groom or even carry it. The amount of contact they achieve varies with species – mother baboons and rhesus macaques protect their babies by swivelling round on their heels, keeping a shoulder between infant and aunt, or moving away from her or threatening her if rank differences between them are great. Some baboon mothers seemed to be considerably harassed, and avoided other females up to ten times more often than usual in the first two or three weeks after birth (Rowell, Din, and Omar, 1968). In contrast, langur mothers (Jay, 1963) permit full access to their infant from the beginning, and will even encourage another female to take the infant by leaning back as she approaches, though they never actually hand it over. Related to this difference, langur mothers can always retrieve their infant whenever they want it from any other female. In baboon and macaque groups, mothers have difficulty in getting their babies back from females higher ranking than themselves and are unwilling to let such females have their infants. Lower-ranking females are more readily permitted, but may be hesitant to take the infant for fear of reprisals. As we have already noted, own children and unrelated adult female 'best friends' are usually allowed to handle the baby from the beginning. Occasional female rhesus and baboons seem to be almost pathological 'baby thieves' – this can occur in the wild (Taylor, 1972) but is perhaps more usually a result of the incomplete nature of the social group in captivity. A high-ranking female baboon would steal new infants even while her own was only a few weeks old, and as mentioned earlier a high-ranking adolescent female rhesus grossly modified the behaviour of two mothers in her group by repeatedly stealing their infants and playing roughly with them whenever she had the opportunity (Rowell, Hinde, and Spencer-Booth, 1964). Interest in infants is not always so great: In Sykes' monkeys, vervets, and talapoins, adult females limited their attentions to rather cursory inspection of the infant (in caged groups) but adolescent and nulliparous young adults were more persistent; probably aunt behaviour is mainly restricted to older siblings in these guenons, in natural groups. In wild vervet groups also, aunt behaviour is largely restricted to sub-adult females (Lancaster, 1972). Spencer-Booth (1968) found that while females about two years older than the infant were the most active aunts, females who had had young of their own showed least interest, especially if they themselves had a young infant at the time. Interaction with older females is by no means always friendly: Mothers with their own infants were most frequently aggressive to other infants in Spencer-Booth's study, especially when the infants were more than four months old, and at Ishasha aggressive responses, usually tweaking or pulling roughly at infants, were more commonly recorded than friendly behaviour from adult females to infants. At Gombe Stream, adult females would lean over the mother's shoulder to threaten infants which their mothers were weaning – in both cases it was hard to avoid the impression that the females were jealous of the mother's attentions to her own infant (Ransom and Rowell, 1972). Females also behave differently towards each infant depending on their relationship with its mother – rhesus females have been observed to break up rough play between infants, threatening one to rescue that of a best friend. Bonnet macaque females would mother an infant whose mother was removed from the group, but ignored a strange infant placed in their cage. Such aggressiveness is extremely rare before the third month. The natal coat of all catarrhines is different in texture and sometimes strikingly different in colour from that of the adults (the grey langur has brown infants, the agouti baboon has black infants, the black colobus has white infants, and so on). In addition the skin of dark-skinned species is pink at birth, and there may be other special characteristics like the centre parting of the new-born rhesus monkey or the red cap of the patas (note that it is the top of the head, the most visible part of an infant at its mother's breast, which has the most obvious infant characteristics). It seems likely that the distinctive appearance serves as a protective signal reducing aggression in conspecifics; in most monkeys the natal coat begins to be replaced around the end of the second month, and aggression towards infants begins to be noticed shortly afterwards.

Infants clearly distinguish between aunts and their own mothers from the time they are seen out of contact with their mothers. They are reluctant to cling to another female and will often try to escape her and return to the mother. If the mother is removed from the group, on the other hand, an infant rhesus will actively solicit attentions of aunts. (Pigtail infants will only do this to an aunt with which they have a particularly good relationship before the mother is removed.) Although it would seem possible that an infant would be successfully adopted if its mother died, there is as yet no evidence for this having occurred in the wild: orphan chimpanzees were adopted by their older siblings (van Lawick-Goodall, 1968), which were too young to be able to give them adequate care. Adoptions have been recorded in several species in captivity.

Males

There is a great deal of variation in the extent to which male monkeys interact with infants. At one extreme, in most of the small New World monkeys that live in small family groups, the male carries his infant for most of the time, handing it over to the female only for feeding. At the other, male patas monkeys, Sykes' monkeys, and most rhesus monkeys ignore infants as far as it is possible. There is a lot of variation even within a genus as closely related as the macaques: barbary males spend a lot of time playing with infants and carrying them (Deag and Crook, in press); Japanese males show considerable paternal behaviour to year-old infants when their mothers give birth to the new seasons' cohort (Itani, 1959); bonnet males are solicitous of infants whose mothers have been removed from the group (Kaufman and Rosenblum, 1967). In rhesus only the occasional male shows interest in infants, and pigtails also interact very occasionally with infants.

Protection of infants has been fairly often recorded, both in the form of attacking or threatening potential enemies (captive males of many species become more aggressive when females in their cage have infants) and in picking up stray infants or even taking from the mother to carry them when the troop flees.

Several studies have found interaction between adult males

and older infants, rather than new borns. A captive male baboon showed only cursory inspection of infants until they began to leave their mothers with some frequency at about six weeks, and then interacted quite extensively with male infants, less with females, for about two months (Rowell, Din and Omar, 1968). Chalmers (1968b) saw adult male white-cheeked mangabeys carry infants in their second six months quite frequently; Itani found a big increase in male–infant interaction in the next birth season when the infants were about a year old, including carrying, grooming, and associating. Young male langurs approach adult males, mount and embrace them (Jay, 1963), but this is perhaps typical of juveniles rather than infants; langur males ignore young infants.

Males of several species have been seen to use infants in interactions among themselves. Descriptions all suggest that a male carrying a young infant reduces the likelihood of other males attacking him – thus a harassed male baboon will snatch up a passing infant and hug it, lipsmacking over its head at the aggressor males, while barbary apes will pick up an infant and start to carry it on the back before approaching a more dominant male.

Development of individual relationships between males and infants have been described in two studies of baboons in the field – that of Kummer (1968a and b) on the hamadryas, and Ransom and Ransom (in press) on the olive baboon at Gombe Stream. It would, however, be premature to suggest that such relationships are peculiar to baboons, and more probable that the unusually good observation conditions in these two studies made the discovery possible.

In hamadryas the relationship between young males and infants seems to form the basis of the specie's typical social organization. Subadult or large juvenile males carry infants away from their mothers and mother and play with them from the time when the infants begin to leave the mothers. A young male forms the centre of infant play groups; they run to him when frightened, and try to gain his support in threatening their infant aggressor. The infants compete for the male's support in such interactions, and he is to each a somewhat ambiguous figure because as well as protecting he will at times threaten the infant in support of another. During the weaning period the infants stay with the male for progressively longer periods, and early in their second year the females have formed a permanent relationship with one young male; in many ways he acts towards his young 'consort' as if he were her mother, but also attacks her if she strays, in the way adult males maintain their groups of females intact. This relationship seems to mature into that of adult male and female, the female using her male in threatening other females in the same way as she used to gain protection from other infants.

Ransom identified four ways in which a bond between an adult male and an infant might develop: Some males had strong affectional bonds with individual older females. When the female gave birth the male paid great attention to the infant (which was in fact very probably his own offspring), carrying and grooming it as part of his interaction with the mother. The second type of relationship was an intensification of the usual protective role of the male, in the instance Ransom and Ransom give as an example, for an infant whose mother seemed unusually uninterested in protecting it herself. She would leave her daughter for long periods and the infant would be carried by one male, the relationship developing first in the context of a predation threat from chimpanzees. The male had no bond with the mother, and though he was often close to her incidentally to his interaction with the infant, this stopped as soon as the infant died.

The third type of relationship was rather similar to that described for hamadryas baboons: young adult or subadult males would kidnap the infants of young subordinate females who had no bond with an adult male themselves and were not able to protect and retain their infants because of their low social status. It seems possible that an adult male–female bond could develop out of such a relationship – a female infant was seen to spend increasing amounts of time with her juvenile male and to run to him for protection or comfort. Again the mother had no special relationship with the male.

The fourth type of relationship was another extension of a

widely reported male–infant interaction pattern, the use of the infant in tense interactions with males. Males chose particular infants for these situations. One was that of a high-ranking female who had newly joined the group and the male was apparently trying to gain rank by associating with this female and her infant. Another chose an older male juvenile who seemed to enjoy the situation so much he would attempt to involve his male in tense interactions with other males; his adult male however began to reject him in favour of new-born black infants which were presumably more effective in such situations.

Conclusions

In analysing the factors affecting social development of monkeys we have used a process of fragmentation, so we must conclude by returning to look again at the normal social environment as a whole.

Although we have considered each class of other animals separately, in overview the most striking aspect is the diversity of interaction possible with each class, and the extent to which their functions overlap – perhaps the only function exclusive to one class of companion is the provision of milk by the mother. Hence there is a lot of flexibility in the system, and it is striking how, in experiments using various degrees of social deprivation, the infants have been able to compensate for the absence of one type of companion by extending their relationship with another (for example, infants reared together but without mothers clung to each other instead).

Infants interact, not with classes of animals, but with individuals. Bowlby (1969) has stressed that the formation of individual relationships is in itself an essential part of normal development in children, almost regardless of with whom the bonds are formed. That this is true for monkeys is suggested by an experiment in which four infants were rotated between mothers every two weeks. The mothers showed apparently normal maternal behaviour, but the infants showed more 'disturbance' behaviour, and at three years they tended to be more dominant (more aggressive?) than animals reared by their own mothers for the whole time (Harlow and Harlow, 1969).

For the most part I have been discussing the infant's interactions as if it interacted with only one animal at a time, but this is of course also an oversimplification. Tripartite interactions involving infants have been discussed by Kummer (1967). He defined these as interactions in which 'three individuals simultaneously interact in three essentially different roles and each of them aims its behavior at both of its partners'. In hamadryas baboons social development could be described in terms of tripartite interactions to a large extent. The infant is first defended by its mother; later it transfers the role of protector to a young male, and the relationship is complicated because the infant competes with several others for the protection of the male; while in maturity the relationships between 'harems', and within them, are again most frequently in the pattern of protector, protected, and antagonist, the baboons manoeuvring to achieve the pattern. Complex interactions of this type are much more commonly seen when watching monkey groups than one would perhaps realize from the literature, since most writers have taken their main task as that of simplification in order to convey their information clearly.

Often other animals affect the infant less by their direct interaction with it than by the indirect effect they have on other relationships. Adult females in a rhesus group have an important effect on the restrictiveness of the mother towards her infant and so modify its early experience; this effect will vary with the relative rank of the mother and the other female, and the relationship of the two females may be modified in turn by the arrival of the infant. An adult male mangabey interacted very little with young infants born into his group (caged) or their mothers, but his presence apparently reassured the mothers and allowed them to be relaxed about their infants. When the male died the females restricted their infants more, at least for a time. This is probably a general indirect effect on infants of adult males, even in species where direct male–infant interaction is rare.

8 Behaviour Between Adults

In this chapter I shall discuss some models of social interaction systems among adults. We shall first consider two which have been very widely used in primate studies, hierarchy and territory. (It is surely a comment on our own species that we have attempted to explain the behaviour of other species almost entirely in terms of concepts defined by aggression.) Then we shall consider the role model of social organization which is currently being developed in relation to monkeys, and finally I shall introduce a couple of other concepts which I predict will prove valuable in future analyses of monkey social organization, although there is very little information about them at present.

Hierarchy

Hierarchy was first studied in animals in the form of the peck-order in flocks of domestic hens, and concerned the predictability of the direction of aggressive encounters among individuals. In primate studies the concept immediately widened in scope. An animal's dominance–subordinance relationships were regarded as the most important social data, and hierarchy was effectively synonymous with social organization in primates. Dominance was usually defined, not in terms of aggressive encounters, but as 'priority of access to desirable objects'. Since desirable objects included oestrous females, dominance would imply increased numbers of descendants for males, and so had important selective advantage.

Dominance relationships were first studied in captive animals, and later in free-ranging but fed groups of macaques and baboons. In the last ten years a large number of field studies on undisturbed groups were reported, and an interesting dichotomy began to emerge, because such studies rarely used the 'fundamental' concept of hierarchy in explaining the social organization observed, or occasionally paid lip-service by stating that hierarchy was 'latent', that is, so well established that it was never seen expressed (an idea which is difficult to distinguish from the hierarchy simply not being there). Recently several people have begun to question the fundamental position of hierarchy in social organization, notably Gartlan (1968) and Bernstein (1970), who also review the use of the concept in earlier literature.

Most caged groups of monkeys have an obvious hierarchy. This was certainly true of the captive group of baboons described in chapter 4. The question was, how important was their hierarchy in their social organization. That is to say, how much of their behaviour could usefully be described in terms of the hierarchy, and how predictable was the outcome of various types of interaction on the basis of hierarchy (Rowell, 1966b). To answer this question, an analysis was made of the distribution of about twenty behaviour patterns. The ranking of the animals according to the frequency with which they performed or received each piece of behaviour was compared with the ranking in the apparent hierarchy obvious to the observer. The consistency with which each behaviour was directed up or down the hierarchy, and its relative frequency in different age- and sex-groupings was also investigated.

Two classes of interaction were recognized, those which had the pattern 'animal approaches – other retreats' and the 'friendly' or 'approach–approach' patterns. I think it is pretty obvious that only the former can be used to infer dominance relationships directly, though one might then find correlation between such relationships and the pattern of friendly interaction. The best agreement with apparent rank came from the non-agonistic approach–retreat interactions, like supplanting, or avoiding an apparently neutral approach (this was by far the most frequent sort of approach–retreat interaction too). Threat, chases, making aggressive contact, and other agonistic behaviour patterns were less well correlated.

Most friendly behaviour was only slightly related, in frequency or direction consistency to the apparent ranking, not well enough to have much predictive value. Some friendly patterns had a spuriously high correlation because they were often directed by one sex to the other: for instance, presenting the hindquarters was frequently directed by the females to the adult male, and so up the hierarchy. An equal number of presentings occurred between females, however, and among them it was directed equally often up and down the hierarchy.

Since behaviour patterns were not entirely consistent in direction, even when they were well correlated with the apparent hierarchy, we can see that the hierarchy is not static, but has some inherent flexibility – it is potentially capable of being altered as conditions change. Probably changes begin in areas of behaviour only slightly correlated with rank, which have little effect in themselves, but which could make a direction reversal in a better correlated behaviour a little more likely. In this way the beginning of a rank reversal would probably pass unnoticed, and adjustments of rank as animals age or mature would take place with minimal disruption. An example of this sort of change occurred in the baboon group. One juvenile female scored a higher rank than expected in several measures, notably in the proportion of her interactions which she initiated herself. In the year following the study quoted, she rose at least one place in apparent ranking. Thus the hierarchy can change as a result of new experiences of the animals involved. It is in fact the expression of the current position in a continuous learning process. Every interaction between two individuals will tend to reinforce or extinguish their dominance relationship produced by all the interactions preceding it. It follows that the first introduction of two animals strange to each other will have a disproportionately large effect on their future relationship, so that a 'stable' hierarchy is formed very rapidly if a group of strange monkeys are put together in a cage. As in other learning processes, the response learned will depend on the number, frequency and pattern of 'trials', and will be capable of extinction or reversal. And again as in other learning situations, a test run will itself have an effect – for example the classic test for dominance of offering food to a pair of animals (the one which takes it is dominant) must strengthen the learned response to this situation – dominant takes, subordinate ignores – with each performance.

Monkeys can also learn by observing a situation in which they are not themselves involved, so that the relationship between two animals is also affected by the interactions of each with other members of the group. (It was for this reason the total frequency with which each animal performed each behaviour pattern, regardless of its partner, was expected to be a useful measure in describing the coherence of social organization in the baboon group mentioned above.)

The much more extensive study of hierarchy by Bernstein (1970) also brought out the complexity of the factors involved. He observed six species of monkeys, in groups living in large enclosures. The frequency and direction consistency of three behaviour patterns commonly regarded as being dominance-related were studied: aggression, mounting, and being groomed. Five of his species were in the macaque–baboon–mangabey group, and all these showed stable hierarchical relationships in agonistic behaviour over several months. The sixth, a guenon, did not show a stable hierarchy on this measure, pairs of animals reversing their relationship several times over a year. Hierarchy based on mounting relationships was rather less stable than that from agonistic encounters, while grooming relationships were essentially reciprocal (non-directional), and could be predicted better from an animal's age and sex than in terms of hierarchy. Most importantly, Bernstein did not find significant correlations between the hierarchies obtained from the three measures in any group. He concluded that the three response relationships were not derived from a single social mechanism, but were independently determined, and not necessarily determined the same way in different species. None of the behaviour patterns was necessarily predictive of other social relationships within the group.

In other words the idea of hierarchy as a single basic underlying factor in monkey social organization does not stand up to critical, quantitative analysis.

Several lines of evidence from the baboon study suggested that hierarchy was maintained by the behaviour of low-ranking animals – that is, we are dealing not with a dominance hierarchy but a subordinance hierarchy. Far from his dominance being the important social character of an animal, it appeared to be what was 'left over' after his degree of subordination was accounted for. For example, the best correlation with apparent rank was given by the frequency with which an animal initiated interactions which were 'misinterpreted' by its partner: subordinate animals seemed to be very careful about their choice of partner, and started, as far as possible, only interactions which would follow a predictable pattern, while high-ranking animals were much less cautious. In fact, this should have been the expected pattern, because the outcome of an approach–retreat interaction must, when you think about it, ultimately be decided by the behaviour of the potential retreater – you can't chase an animal which won't flee. That subordinance rather than dominance should be the identifiable behavioural phenomenon also fits well with what we know of the physiological concomitants of establishing a hierarchy. As described at the end of chapter 4 low-ranking animals are under physiological stress, having enlarged and hyperactive adrenals. Low-ranking animals are also those which succumb most readily to a variety of diseases, many of which have been described as stress-related, most typically an assortment of diarrhoeal diseases. It might be useful to think of the type of chronic intense subordinate behaviour that is seen in the low-ranking animals of very clear-cut hierarchies (which probably actually maintains such hierarchies) as another stress symptom, comparable to the adrenal changes and susceptibility to disease. This would conceive the hierarchy itself as a sort of pathological response of a social system to extremely stressful environmental conditions.

It has usually been assumed that the function of hierarchy is to reduce aggression (Collias, 1953), but in fact the observable correlation is between clear hierarchies and high levels of aggression. It is always possible, of course, that in such groups the level of aggression is being lowered from something still higher, but this would be very difficult to prove.

If you put strange monkeys together, there is a lot of fighting, and then a hierarchy emerges. The assumption has been that the hierarchy replaces the fighting, but it would perhaps be more economical to say simply that it has been caused by the fighting. Rose, Holaday and Bernstein (1971) have recently found a correlation between testosterone levels and aggressiveness in an all-male group of rhesus monkeys. This meant a general correlation between high testosterone levels and high rank, but not a close one since very high-ranking animals were not very aggressive, and had lower testosterone levels. Bernstein suggests that the output of the hormone is determined by behaviour rather than vice versa (via the pituitary–brainstem link) and this is certainly borne out by the fact that even his low-ranking socially-active animals had much higher testosterone concentrations in their blood than has previously been reported from males living separately in laboratory cages. This is, I think, the first positive correlation between high rank and a measured physiological change.

So far we have discussed hierarchial relationships simply in terms of social behaviour, but dominance has frequently been regarded as a generally favourable quality, conferring a selective advantage on its possessor. Since we have seen that even between different types of social behaviour there is little or no correlation of hierarchies, we may be sceptical of a suggestion that hierarchical position will predict relationships with the rest of the environment. None the less, high rank has often been associated with 'leadership' in various forms. There is no logical reason to expect a relationship between high rank as seen in approach–retreat relationships and 'leadership', because group behaviour is rarely if ever determined by coercion. Thus there is no way a baboon can force its troop to follow a particular route, so there is no reason to expect high-ranking animals to lead in a literal sense in the day's march. Nor are higher-ranking animals necessarily better at finding food or discovering and avoiding danger.

Territory

The concept of territory has been used in discussing the relations between groups of monkeys, rather than between individuals. It was first developed by ornithologists, describing the defence of a small area by a cock bird, or by a pair of birds, during the breeding season. A territory, then, is a defended area. It is useful to distinguish territory from home range, which is the whole area in which an animal lives, but which is usually shared with other animals. A territory is recognized by the sharp change of behaviour of its owner at its border – within the territory the owner is confident and aggressive, outside it he is timid and defensive towards strangers. It is also practical, then, to distinguish between aggressiveness to strangers which only occurs in relation to a defined piece of topography, and aggression on sight, or defence of a 'personal space'. The latter is perhaps more common than territorial behaviour in larger mammals, but has often been taken to be territoriality.

Passerine birds announce their territory by singing. By analogy it has often been assumed that loud noises made by mammals, and more especially scent marking, which is often accompanied by elaborate special behaviour, are also means of demarcating territories. Ralls (1971) has shown that this is not a likely interpretation of scent marking in some of the small antelopes and other species, and anyone who has kept a dog knows that the typical leg-lifting scent-marking pattern is not confined to the area the dog will defend – it occurs in totally strange surroundings, and the only place the dog does *not* scent mark is his own house, which he will defend from other dogs.

Territory has repeatedly been linked to hierarchy, since both have been regarded as a system of competition within the species which results in selection of the fittest individuals for breeding purposes. Where animals maintain true territories, however, their hierarchical relationships change with place, the local territory owner being always dominant. Thus birds coming to a feeding station have an obvious peck-order, but the highest-ranking bird is the one in whose territory the feeder happens to have been placed, and a different peck-order would develop in a different site (Brian, 1949). It has been suggested that hierarchy and territory were alternative ways of harnessing male aggression in primate societies – a hypothesis which is rather ahead of our current knowledge, though it is noteworthy that one of the clearest accounts of territorial behaviour in primates, Struhsaker's (1967b), of the vervet monkeys of Amboseli, concerned groups which were also hierarchically organized.

Confusion on this subject has arisen in primate studies because Carpenter, one of the first of the modern field-workers, used the term extensively but with an unusual meaning: unlike other zoologists, he did not distinguish territoriality from home range (e.g. Carpenter, 1958). In some species the two are indeed the same space, as in the howler, which Carpenter described.

The least equivocal accounts of territorial behaviour, which would satisfy the most stringent definition of the term, are in primates which live in family groups, a single pair with their immature offspring. Thus gibbons (Elefson, 1968) and titis (*Callicebus moloch*) (Mason, 1968) both live in small areas with little overlap between the range of neighbouring families. In both species, regular encounters occur in the 'no-man's land' in which the two families exchange threats, chase each other, and exhibit general excitement (hair on end, etc.), after which each retires towards the centre of its own area and feeds peacefully.

Territorial behaviour, in which groups meeting at specific places presumed to be boundaries lined up and threatened each other, has been described in vervet monkeys by Struhsaker (1967b) and at one of his study sites by Gartlan (1966). Encounters between groups which included some excited display by the adult males have been tentatively described as territorial in two other, forest, guenons (Aldritch–Blake, 1970; Struhsaker, 1969), but at the moment these stand as rather isolated examples among the Old World monkeys, while several observers have commented on the rarity with which their groups encounter others at all, or the peacefulness of meetings of troops at waterholes, etc. There may also be approach–retreat relationships between troops of monkeys which are not related to their position, but usually to the relative size of the troops (Southwick, 1962); and Kummer (1968a and b) describes a battle royal between bands of hamadryas over bait he provided.

Many species of monkeys have a highly distinctive loud

noise, often made only by adult males. In some there has been anatomical specialization to make even louder noise, large air sacs and associated bony structures to support them, making resonating chambers for some very remarkable noises indeed – the howler monkey, the gibbons, and colobus monkeys being the most extreme examples. These loud noises are obviously for communication between groups, and not with other group members a few metres away, and it is probably significant that they are best developed in species whose diet and anatomy restricts them to habitats with the densest vegetation and poorest visibility.

Carpenter described tremendous bouts of 'singing' by the adult males when two howler groups come close together, and assumed that in these extremely unaggressive monkeys the males 'fight' with their voices rather than with more conventional weapons, so that the 'singing' was in fact competitive, and a means of establishing territorial boundaries. Howlers also sing at about dawn, and the boundary singing is actually rather rare, considering the small size of their ranges.

Marler (1969) studied a population of black and white colobus monkeys. The male colobus also sing at about dawn, whenever there is a disturbance in their area, and sometimes apparently for no reason at all; singing is infectious, neighbouring males taking up the chorus when one male starts to sing. Marler also observed a few encounters between groups of colobus – again, as in the howlers, they were remarkably rare considering the population density in that forest. During the encounters the adult males bounced in the branches, threatened, and chased each other, but they did not sing. Thus the singing of colobus monkeys is not an agonistic activity, nor is it concerned with territoriality as such. Rather it has a more general communicative function, allowing the whole of a local population to maintain contact and follow each other's whereabouts so that each small group of a dozen animals or less has wider social relationships than their exchanges among themselves might suggest. The singing probably serves as a spacing mechanism and accounts for the relative rarity of face-to-face contact between troops, and yet it is not aversive. The bouncing and threatening between adult males when the troops do meet, on the other hand, probably is territorial defence, though they possibly behave in this way simply because they have come close to each other, invading each other's 'individual distance'.

Perhaps Marler's analysis for colobus might prove a better explanation of the behaviour of howler monkeys also; gibbons, however, howl during face-to-face encounters before and between chases and other agonistic behaviour – so here the association between the noise and territorial defence is much closer.

In summary, territorial behaviour in the strict definition of defence of a particular piece of ground is not very frequent among primates. It is one of a whole range of ways in which monkeys live and maintain social relationships with the rest of the small group of which they are a part.

Roles

In an effort to find an alternative to hierarchy as a descriptive model for primate social organization, several workers have attempted to describe the behaviour of monkey groups in terms of social roles. A role may be defined as a consistent pattern of response shown by particular members of society in specified situations (Bernstein, 1966). Bernstein and Sharpe (1966) described a breeding group of rhesus monkeys in terms of behaviour clusters characteristic of different age and sex classes, and this proved at least as illuminating as an analysis of their hierarchical relationships, which were not manifest in much of their interaction anyway. Gartlan (1968) showed that members of a wild vervet group similarly had a different set of behaviour patterns depending on their age and sex.

Unless the concept of role goes rather further than this, however, it does not seem to add a great deal to understanding social organization beyond a preliminary description. If we are to deal simply with gross age–sex classes and single behaviour patterns, it would be sufficient simply to state that adult males do this and juvenile females do that, without invoking any theoretical concept. Such a treatment succeeds in the original intention of describing the behaviour of monkeys with attention to its complexity and without referring everything to relatively infrequent competitive or agonistic behaviour.

The role concept becomes valuable if it is possible to distinguish *clusters* of behaviour typical of *individuals* more narrowly identified than just by age and sex. Such an analysis was made by Bernstein (1966) on a group of capuchins (*Cebus albifrons*). He identified a particular adult male as 'control animal' using a series of tests introducing new and somewhat alarming situations to the group. This male controlled intra-group disturbances and led the responses of the group to external challenges. When alarmed, other members of the group approached the control animal even if this meant also approaching the source of alarm.

This behaviour has often been identified as that of the 'dominant male' or the 'alpha male' in baboon and macaque groups, but in the capuchin group there was no apparent hierarchical structuring. 'Leader male' has also been used, but implies travelling first in group progressions, which such animals seldom do. Bernstein therefore suggests that the 'control animal' role can be separated from concepts of dominance – nor is there any *a priori* reason why it should be a male role in all cases. This study is a promising start to the development of the role concept for describing monkey societies, but as Bernstein says, the control animal is the most striking role, and there is so far no progress in describing more subtle roles.

Kinship

Because primates grow and mature so slowly, there are very few studies in which it has been possible to examine the effect of kinship on interaction patterns, but such evidence as we have suggests that it may be extremely important. As usual the macaques are best known in this. Information has come from the long-term studies of Japanese macaques (Kawai, 1958; Kawamura, 1958) and of rhesus macaques on Cayo Santiago (Koford, 1963; Sade, 1965, 1967). In both these species, groups are organized in matrilines, mothers, daughters, and grand-daughters maintaining close social relationships. Within a matriline, rank orders run from mother to youngest daughter, through successively older daughters. The daughters' own children are interpolated in the series, again the youngest being next in rank to the mother, followed by older siblings. Whole matrilines have ranking relationships with each other. Males tend to leave the immediate proximity of their maternal relations, but their interaction patterns are still influenced by them: Grooming usually occurs within matrilines, rarely across them. Mating on the other hand is rather more frequent between animals of different matrilines, and sons very rarely mate with their mothers. The rank of a male depends on that of his mother so long as he remains in his natal group, but they quite often move to other troops and their rank may then change abruptly. Kinship may still operate during and after a male's transfer, however – Wilson and Boelkins (1971) describe transferring males being introduced to a new group by 'sponsor' males, which were often identified as elder brothers who had transferred into the new group in a previous season. Sade tells a vivid story of a plastic imitation apple which he offered to one of his rhesus groups on Cayo Santiago, and which, as he said, demonstrated in a few minutes the relationships it had taken him many months of patient observation to work out. The apple was taken by the alpha male of the group, inspected and discarded; it was grabbed by the senior female of the highest ranking matriline, passed rapidly from hand to hand, mother to daughter, then was taken by the senior female of the second matriline and inspected in order by all her descendants, and so on through the group until it was finally discarded by the low-ranking juvenile males.

In other species kinship has rarely been discussed beyond the context of mother–infant interaction and the role of siblings in caring for infants, and in view of the picture from the macaques just described, it has undoubtedly been underestimated as a determinant of interaction patterns between adults. Tantalizing hints of this can be found here and there: van Lawick-Goodall (1968) was able to identify a whole series of offspring of one female chimpanzee. Her adult and subadult sons lived for the most part in male groups, but would occasionally stay with their mother for a short time; when they met their affectionate behaviour towards her and their younger siblings showed clearly that they still maintained a special family relationship. Ransom (1971) remarked on grooming relationships between adult baboons which had no obvious basis in the immediate situation and were most probably based on kinship.

As we noted in chapter 4, one of the major differences between artificially formed captive groups and wild troops is that the normal kinship network is inevitably lacking for several years. Adult females are most involved in these relationships in the wild, and it is this class of animal which suffers most from the trauma of capture and finds relationships in a new group most difficult to establish. In captivity studies also, then, the effects of kinship have usually been ignored or greatly underestimated, though probably a major part of the differences between social behaviour in wild and cage environments may well be due to this single factor. I would predict that kinship will prove to be the most important key to understanding social organization of natural groups of monkeys, and that its study will well repay the rather formidable practical problems involved in maintaining any really long-term study.

If on the other hand the main purpose of primate studies is regarded as being to provide insight into human social organization, newly formed artificial groups probably have much to offer. They show many parallels with modern urban human groups, not least in the high level of stress they apparently impose on their members.

Age

If kinship has been relatively little studied, it is at least generally recognized to be a theoretically important factor. Though age has of course been rather carefully considered as a variable in developmental studies of social behaviour in infants and juveniles, there has been a general assumption that social development stops at puberty in female monkeys, and continues only a little longer in males, until they reach full 'social maturity' as adult males. That this is not so was brought home to me in a study of a small group of captive adult white-cheeked mangabeys (Chalmers and Rowell, 1971). There were only four animals, and it so happened that the male and one of the females were very obviously older than the other two females. There were notable similarities between the two old and the two young animals which cut across divisions of sex and rank. The two older animals initiated relatively few interactions, but were very frequently approached for social interaction and groomed by the younger ones; other differences, together with these, suggested 'deference behaviour' by the younger towards the older animals.

Some differences between adults which were related to age have already been discussed. In chapter 6, we saw that old females of several species showed more intense and longer bouts of oestrous behaviour than did young ones, and that old female baboons behaved differently towards their new-born infants, providing them with a different social milieu from young mothers. Kummer discussed social relationships of male hamadryas baboons in terms of a developmental process which started with the juvenile and continued into extreme old age, with the male taking a series of very different roles in relation to his female group and the band as a whole during his maturity (*see* chapter 3).

One interesting possibility is that many of the traits that field-workers have taken to be individual characteristics may in fact be typical age-specific patterns. This would considerably affect many existing theories: supposing, for example, that the role of the alpha male, or of the control animal, was one that most males assumed for a short period some time in the middle years of adulthood, and not a rare characteristic of a few animals. Similarly, copulation frequency in males of other species besides hamadryas could again be a function of age, and reproductive success be much less variable than has been thought. In that case, ideas about selection pressures for 'dominance behaviour' would have to be modified. There would of course still be natural selection pressures on males ensuring that only the 'fittest' reproduced, but rather than selecting for specific characters of dominance or lecherousness, for the most part there would be a more general selection for survival until the age when such behaviour usually develops. One implication is that there might be rather less difference between individual monkeys than is our current impression, but that each individual has more varied roles when its life is considered as a whole. The point is taken up again in the final chapter.

The main problem of investigating age as a determinant of adult behaviour is of course that of finding out how old monkeys are; general appearance is a poor indicator, since sick animals usually look old to us. Tooth wear is a reliable indicator of relative age of animals from the same environment. (In species where the males show tension yawning, like baboons, it is usually possible to rank them for age on these glimpses of tooth wear even in the field.) Those of us who work on caged animals should at least be able to rank our animals roughly by age, once it has been established as a potentially important variable.

9 Some Considerations of Function

So far we have considered mechanisms – how monkeys interact, how their social groups are organized. A zoologist however must always return to the question of selective advantage – that is, one asks 'why', in a philosophically limited sense. When I am asked why monkeys live in groups, the first answer that comes to mind is 'because they like each other's company' – and that is probably a better answer than many more seemingly sophisticated answers that have been offered like 'sexual attraction'. It is so very obvious that monkeys enjoy being together that we take it for granted. But pleasure like every other phenomenon of life is subject to, and the result of, evolutionary pressure – we enjoy a thing because our ancestors survived better and left more viable offspring than their relations who did not enjoy (and so seek) comparable stimuli.

Because monkeys clearly enjoy each other's company very much, we may infer that sociability is very important to their survival, or at least has been in the recent past. Field and experimental work on animal associations has occupied zoologists for a long time and there is a body of theory on the subject. Monkeys are not good subjects for this sort of work because they are not generally numerous or easy to observe in the wild, and are too expensive and difficult to maintain in experimental conditions. We assume that general propositions derived from other animals will be valid for primates too, but it should be realized that there is no direct evidence for many of the functions of monkey social behaviour which are generally attributed to it. Some of these generally assumed functions of social behaviour are discussed below.

Conditioning of the medium

Protozoa, fish, and other aquatic animals survive better in groups, or even in the water in which others of their species lived; apparently chemical changes in the composition of the water due to the excretions of the other animals improve it as a habitat. This is social behaviour in a much wider sense than we have used it in this book, since conditioning and benefiting animals need never meet. This concept is more difficult to apply directly to land animals, but a parallel might be in establishing trails leading to good food or water sources, salt licks and so on. The regular pathways of monkeys through the trees are generally free of treacherous dead branches for example. A rather longer-term way in which monkeys improved their habitat for their descendants was discovered by Gartlan and Jackson on Lolui Island in Lake Victoria, where vervet monkeys lived at the boundary of forest and grassland. They fed mainly on forest tree fruits, but liked to go out into the grassland to sit on termite mounds to rest. Termite mounds formed the focus points of recolonization of the grass by forest, and the first colonizers were the trees whose fruit was preferred by the monkeys, the seeds being deposited there in their faeces or carried undigested in cheek pouches.

Efficient foraging

By foraging in groups a population of animals may obtain a larger share of available food than another species foraging singly since a chance find by one animal can be used by the whole group. To work, this needs the right sort of food source, well hidden but providing plenty when discovered – fruit trees in forest, or swarms of insects would be good examples. It works best where animals are socially alert, and can imitate their fellows' foraging behaviour or at least show social facilitation of foraging; these are criteria on which monkeys generally rank exceptionally highly.

Defence against predators

Several pairs of eyes and ears are better than one, and it is rarely possible to approach a large group of monkeys without the alarm being given. Occasional solitary animals are said to be easier to take by surprise, though since they are usually adult males which are less easily alarmed than other classes of animals, this is not necessarily direct evidence for the protection provided by groups. Some adult prey animals may be sufficiently well armed literally to defend juveniles from predators. Baboon males are usually cited for this, although I myself have only seen them lead the retreat. And there are other tactics available once a predator has found the group. A predator needs to concentrate on a single animal to mount a successful attack, and it is very difficult to single out one in a fast-moving crowd, as one finds when trying to net animals from a group. Mock attacks, and screaming by all the troop further confuse the attacker.

For schooling fish, it has been suggested that social behaviour protects from predation by leaving large areas of sea empty of prey at any one time, and this might also work at low densities for monkeys in forest and even better in sparse open country. On the other hand the predator is not likely to miss a large group of prey if it should find them, nor to lose them again easily. For this mechanism to be effective the groups of prey would have to be sufficiently rarely encountered for it not to be worthwhile the predator establishing a search image for the species (predators, having captured prey, tend to look for another of the same sort, which provides, for example, the selection pressure for polymorphism in some easily captured animals like snails).

The usual strategy of forest monkey groups against predators they discover is to give the appropriate alarm call, different for ground and aerial predators, and then for most of the group to move a short distance into cover while a few individuals mark the source of danger and make noises which comment on its movements to the rest of the group. This strategy has become disastrous with the advent of firearms, and throughout West Africa, where monkeys are a delicacy, most species are moving rapidly towards extinction as sentinels are easily picked off. Species disappear in a definite order, those whose alarm display is most obvious and flight least effective going first. This new selection pressure is probably too heavy for most species to survive (protection of primates is either not attempted or unenforced) but there is some evidence that it is producing drastic changes in behaviour in some places, the alarm displays being reduced (Gartlan, personal communication). Group size is smaller in these areas of heavy hunting, perhaps just because so few animals remain, but possibly also because the old strategy which depended on large groups is no longer effective – to avoid this new overwhelming predator, one must be silent and unobtrusive above all.

Breeding synchrony

Animals may be brought into breeding condition by the sight, sound, or smell of conspecific breeding activity. If the whole of a large group breeds at the same time, predators, which have to live for the rest of the year as well, will only be numerous enough to take a small proportion of the young (*see* chapter 6). Where the season in which young may be successfully fed is limited, it may also be advantageous to synchronize the onset of breeding at the earliest possible moment in the season. This effect is probably not important for most primates, but it would be worth looking for where some breeding synchrony is observed – do young born in mid-season survive better than those born early or late, which would indicate the presence of this sort of selection pressure?

Genetic advantages

Living in groups may be an advantage to the species in a purely genetic sense: if breeding occurs exclusively within small groups the genetic variability of the species will probably increase, so that it is more likely to be able to meet environmental changes with an appropriate pre-adaptation. Again, there is the possibility, in more or less closed breeding groups, for a form of group selection to occur: it is possible for one animal, by its behaviour, to enhance the possibilities of survival of its offspring or its close relations with similar genetic composition (by protecting it from predators or leading it to important food sources, for example). Only in a closely related group could such 'altruistic' behaviour be advantageous – saving other people's offspring does not increase your own contribution to the gene pool. On the other hand recent long-term studies, as we have seen, have been finding more movement of individuals between troops than had been expected.

Changes in the behaviour of our own species occur far more often and more frequently than in other animals because, as Huxley (1942) pointed out, genetic evolution has been largely replaced by cultural evolution. Information is no longer passed to new generations only in genetic material, but also by communication, at first between individuals and later more generally through writing and other media. The evolution of more and more sophisticated social behaviour has eventually led to a new 'breakthrough' in evolution itself.

For a long time group living by other animals was thought to be merely pre-adaptive for this human breakthrough and of immediate advantage to social animals only in the more mechanistic ways suggested above. At the same time imitative learning in monkeys in captivity was well known ('monkey see, monkey do'). In the 1950s however, a group of workers in Japan led by Itani (1958) saw the acquisition of entirely new foraging patterns in troops of Japanese macaques, and because they knew the age and relationships of many individuals, they were able to follow the spread of the new behaviour through the troops. They recognized this process as essentially a part of what Huxley had called cultural evolution, and Kawai (1965) described it as 'pre-cultural' behaviour. Later these workers were able to extend their observations to deliberately introduce new behaviour in some troops, and to analyse some of the factors involved in the learning and transmission processes (Tsumori, 1967). In the transmission of recognition of a new food (sweets) they found, as might be expected, that the lively and inquisitive juveniles learned most quickly. They recognized several routes of diffusion of the new behaviour: mothers learned from their infants, and *their* mothers learned from them; older brothers and sisters learned from their younger siblings; adult males learned from adult females with whom they had special relationships; and some old males, who looked after yearlings when their younger siblings were born, learned directly from these juveniles. Thus diffusion was by no means random, but depended on a clearly defined structure of relationships, and males, being less involved in kinship ties, learned much more slowly than females. In 1953 a female of one troop had the idea of washing sweet potato in the sea before eating it (so removing the sand), and by 1962 all animals under twelve years old did this regularly, but very few over that age. A more difficult pattern, using a 'placer-mining' technique to separate wheat from sand in water was learned most quickly not by the youngest animals, but by young adults whose greater experience perhaps enabled them to grasp the technique more quickly. The proportion of monkeys that would learn to dig up buried peanuts dropped very sharply in age classes over seven years, apparently because older animals simply became uninterested in this type of problem solving; females did consistently better than males at all ages because they paid more attention to the test situation, while males were easily distracted.

It may well be that older animals, and especially males, may be adept at other forms of learning, perhaps something more subtle than research workers have yet considered; but at the moment acquiring new skills seems to be part of the role of socially immature animals so that at first sight they appear to be taking the major part in 'pre-cultural behaviour'. I suggest that the role of older animals in cultural transmission might be that of providing a greatly extended group memory, which would have obvious survival value in conditions of infrequently recurring crises. One of the most striking human characteristics is the survival of some individuals well beyond reproductive age – in the extent to which it occurs the species is almost certainly unique, though there are indications that more than the rare, occasional animal may survive into old age in other species, perhaps hamadryas baboons (*see* chapter 3). In order to be valuable, older animals must be distinguishable from younger ones beyond the stages of reproductive and social maturity. Man has a remarkable series of age changes allowing age to be assessed from appearance with surprising accuracy, most notably the greying of the hair. Gorilla males have a similar greying of the hair on their backs and sides; 'silverback' males are middle aged and older, and groups of gorillas are usually led by a silverback male. No monkey has such obvious identification marks for older animals: perhaps their use of the extended group memory is less well developed, but possibly also obvious recognition signals of age are unnecessary between mutually known individuals while people (and perhaps gorillas ?) communicate also with strangers.

There must also be behavioural distinction: there must be something comparable to 'deference' if the group is to profit from the experience of elders in such situations as choice of feeding route or predator avoidance in unusual circumstances. The mangabey group described in chapter 8 provided a hint that age-directed deference may be found in monkey social structures; this would be a fundamental step in the change from genetic evolution to cultural evolution as the main process of behavioural change. The protection and use of the aged individual's experience would then seem a more likely function of the tightly cohesive permanent groups of primates than the more usually suggested protection of young. Infants are easily replaced with relatively little expenditure of energy, and are in any case successfully reared in less structured animal social systems. More than merely ensuring their survival the group might have the less obvious function of providing the best environment for the exploratory learning which will make some of the juveniles into the valuable memories of the future.

This is speculation; but it is by research which examines the function of social systems of monkeys and other animals that we shall be able to understand fully their mechanisms, and this seems to me an exciting lead to pursue.

References

ALDRICH-BLAKE, F. P. G. (1970), 'Problems of social structure in forest monkeys', in J. H. Cook (ed.), *Social Behaviour of Birds and Mammals*, Academic Press.

ALDRICH-BLAKE, F. P. G., DUNN, T. K., DUNBAR, R. I. M., and HEADLEY, P. M. (1971), 'Observations on baboons, *Papio anubis*, in an arid region in Ethiopia', *Folia Primatologica*, no. 15, pp. 1–35.

ALTMANN, S. A., and ALTMANN, J. (1970), *Baboon Ecology*, S. Karger, Basel.

ANDREW, R. J. (1963a), 'Trends apparent in the evolution of vocalization in the Old World monkeys and apes', *Symp. Zool. Soc.*, London, vol. 10, pp. 89–101.

ANDREW, R. J. (1963b), 'Evolution of facial expression', *Science*, vol. 142, pp. 1034–41.

ANTHONEY, T. R. (1968), 'The ontogeny of greeting, grooming, and sexual motor patterns in captive baboons (super species, *P. cynocephalus*)', *Behaviour*, vol. 31, pp. 358–72.

BALDWIN, J. (1969), 'The ontogeny of social behaviour in squirrel monkeys (*Saimiri sciureus*) in a semi-natural environment', *Folia Primatologica*, vol. 11, pp. 35–79.

BALL, J., and HARTMAN, C. G. (1935), 'Sexual excitability as related to the menstrual cycle in the monkey', *Amer. J. Obstetrics and Gynecology*, vol. 29, pp. 117–19.

BERNSTEIN, I. S. (1966), 'Analysis of a key role in a capuchin (*Cebus albifrons*) group', *Tulane Studies in Zool.*, no. 13, pp. 49–54.

BERNSTEIN, I. S. (1970), 'Primate status hierarchies', in L. A. Rosenblum (ed.), *Primate Behaviour*, Academic Press.

BERNSTEIN, I. S., and SHARPE, L. G. (1966), 'Social roles in a rhesus monkey group', *Behaviour*, vol. 26, pp. 91–104.

BOWLBY, J. (1958), 'The nature of the child's tie to its mother', *Internat. J. Psychonal.*, vol. 39, pp. 350–73.

BOWLBY, J. (1969), *Attachment*, vol. 1 of *Attachment and Loss*, Hogarth and Penguin.

BRAMBLETT, C. A. (1969), 'Non-metric skeletal age changes in the Darajani baboon', *Amer. J. phys. Anthrop.*, vol. 30, pp. 161–71.

BRIAN, A. D. (1949), 'Dominance in the great tit', *Parus major*, *Scot. Naturalist*, vol. 61, pp. 144–55.

Buettner-Janusch, J. (1966), 'A problem in evolutionary systematics: nomenclature and classification of baboons, genus *Papio*', *Folia Primatologica*, vol. 4, pp. 288–308.

Carpenter, C. R. (1964), *Naturalistic Behaviour of Nonhuman Primates*, Pennsylvania State University Press (collection of all earlier papers).

Chalmers, N. R. (1968a), 'Group composition, ecology and daily activities of free-living mangabeys in Uganda', *Folia Primatologica*, vol. 8, pp. 247–62.

Chalmers, N. R. (1968b), 'The social behaviour of free-living mangabeys in Uganda', *Folia Primatologica*, vol. 8, pp. 263–81.

Chalmers, N. R. (1972) 'Comparable aspects of early infant development in some captive cernopithecines' in F. E. Poirier (ed.), *Primate Socialization*, Random House.

Chalmers, N. R., and Rowell, T. E. (1971), 'Behaviour and female reproductive cycles in a captive group of mangabeys', *Folia Primatologica*, vol. 14, pp. 1–14.

Collias, N. E. (1953), 'Social behaviour in animals', *Ecology*, vol. 34, pp. 810–11.

Conaway, C. H., and Koford, C. B. (1965), 'Estrous cycles and mating behaviour in a free ranging herd of rhesus monkeys', *J. Mammal.*, vol. 45, pp. 577–88.

Conaway, C. H., and Sorenson, M. W. (1966), 'Reproduction in tree shrews', in *Comparative Biology of Reproduction in Mammals*, Academic Press.

Crook, J. H. (1966), 'Gelada baboon herd structure and movement: a comparative report', *Symp. Zool. Soc.*, London, no. 18, 237–58.

Crook, J. H., and Gartlan, J. S. (1966), 'Evolution of primate societies', *Nature*, vol. 210, pp. 1200–203.

Daanje, A. (1950), 'On locomotory movements of birds and the intention movements derived from them', *Behaviour*, vol. 3, pp. 48–98.

Darwin, C. (1872), *The Expression of the Emotions in Man and Animals*, Murray.

Deag, J. M., and Crook, J. H. (in press), quoted in J. H. Crook (ed.), 'The socio-ecology of primates', *The Social Behaviour of Birds and Mammals*, Academic Press, 1970.

DeVore, I. (1963), 'Mother–infant relations in free-ranging baboons', in H. L. Reingold (ed.), *Maternal Behavior in Mammals*, Wiley.

DeVore, I. (1964), 'A comparison of the ecology and behavior of monkeys and apes', *Viking Fund Publications in Anthropology*, no. 37, pp. 301–19.

DeVore, I. (1965), 'Male dominance and mating behavior in baboons', in F. A. Beach (ed.), *Sex and Behavior*, Wiley.

Eisenberg, J. (1966), 'The social organizations of mammals', *Handbook Zool.*, vol. 10, no. 7, pp. 1–92, Berlin.

ELEFSON, J. O. (1968), 'Territorial behavior in the common white-handed gibbon, *Hylobates lar*', in P. Jay (ed.), *Primates*, Holt, Rinehart & Winston.

ERIKSON, L. B., REYNOLDS, S. R. M., and DE FEO, V. J. (1960), 'Menstrual irregularities in rhesus monkeys elicited by reserpine administered on selected days of the cycle', *Endocrinology*, vol. 66, pp. 824–31.

FADY, J. C. (1969), 'Les jeux sociant; le compagnon de jeux chez les jeunes. Observation chez *Macaca irus*', *Folia Primatologica*, vol. 11, pp. 134–43.

GARTLAN, J. A. (1966), 'Ecology and behaviour of the vervet monkey', Ph.D. thesis, University of Bristol.

GARTLAN, J. S. (1968), 'Structure and function in primate society', *Folia Primatologica*, vol. 8, pp. 89–120.

GARTLAN, J. S. (1970), 'Preliminary notes on ecology and behaviour of the drill (*Mandrillus leucophaeus*)', in J. Napier and P. Napier (eds.), *Old-World Monkeys: Evolution, Systematics and Behaviour*.

GARTLAN, J. S., and BRAIN, C. K. (1968), 'Ecology and social variability in *Cercopithecus aethiops* and *C. mitis*', in P. Jay (ed.), *Primates*, Holt, Rinehart & Winston.

GILMAN, J., and GILBERT, C. (1946), 'The reproductive cycle of the chacma baboon with special reference to the problems of menstrual irregularities as assessed by the behaviour of the sex skin', *South African Medical Sci.*, vol. 11 (biological supplement), pp. 1–54.

GOY, R. W. (1968), 'Organizing effects of androgen on the behaviour of rhesus monkeys', in R. Michael (ed.), *Endocrinology and Human Behaviour*, Oxford University Press.

HADDOW, A. J. (1952), 'Field and laboratory studies on an African monkey, *Cercopithecus ascanius schmidti*', *Proc. zool. Soc.*, London, vol. 122, pp. 297–394.

HALL, K. R. L. (1962a), 'Numerical data, maintenance activities and locomotion of the wild chacma baboon, *Papio ursinus*, *Proc. zool. Soc.*, London, vol. 139, pp. 181–220.

HALL, K. R. L. (1962b), 'The sexual, agonistic and derived social behaviour patterns of the wild chacma baboon (*P. ursinus*)', *Proc. zool. Soc.*, London, vol. 139, pp. 283–327.

HALL, K. R. L. (1963), 'Variations in the ecology of the chacma baboon (*P. ursinus*)', *Symposium of zool. Soc.*, London, vol. 10, pp. 1–28.

HALL, K. R. L. (1965), 'Behaviour and ecology of the wild Patas monkey, *Erythrocebus patas*, in Uganda', *J. Zool.*, vol. 148, pp. 15–87.

HALL, K. R. L., BOELKINS, C. R., and GOSWELL, M. J. (1965), 'Behaviour of patas monkeys in captivity', *Folia Primatologica*, vol. 3, pp. 22–49.

HALL, K. R. L., and DEVORE, I. (1965), 'Baboon social behaviour', in I. DeVore (ed.), *Primate Behaviour*, Holt, Rinehart & Winston.

HANBY, J. P., ROBERTSON, L. T., and PHOENIX, C. H. (1971), 'The sexual behaviour of a confined troop of Japanese macaques', *Folia Primatologica*, vol. 16, pp. 123–43.

HANSEN, E. W. (1966), 'The development of maternal and infant behavior in the rhesus monkey', *Behaviour*, vol. 27, pp. 107–149.

HARLOW, H. F. (1969), 'Age mate or peer affectional system', in D. S. Lehman, R. A. Hinde, and E. Shore (eds.), *Advances in the Study of Behaviour II*, Academic Press.

HARLOW, H. F., and HARLOW, M. K., (1965), 'The affectional system', in A. M. Schrier, H. F. Harlow, and F. Stollnitz, (eds.), *Behaviour of Non-Human Primates*, vol. 2, Academic Press.

HARLOW, H. F., and HARLOW, M. K. (1969), 'Effects of various mother–infant relationships on rhesus monkey behaviour', in B. M. Foss (ed.), *Determinants of Infant Behaviour IV*, Methuen.

HARRINGTON, J. R. (1971), 'Olfactory communication in *Lemur fulvus*', Ph.D. thesis, Duke University.

HARTMAN, C. G. (1932), 'Studies in the reproduction of the monkey, *M. rhesus*, with special reference to menstruation and pregnancy', *Contributions to Embryology of the Carnegie Institute of Washington*, vol. 23, pp. 1–161.

HEAPE, W. (1900), 'The sexual season of mammals and the relation of pro-oestrum to menstruation'. *Quart. J. Microscopical Sci.*, vol. 44, pp. 1–70.

HEDIGER, H. (1955), *Studies of the Psychology and Behaviour of Captive Animals in Zoos and Circuses*. Butterworth.

HERBERT, J. (1970), 'Hormones and reproductive behaviour in rhesus and talapoin monkeys', *J. Reproduction and Fertility Supplement*, vol. 11, pp. 119–40.

HINDE, R. A. (1969), 'Analysing the roles of the partners in a behavioral interaction – mother–infant relations in rhesus macaques', *Annals of the New York Acad. Sci.*, vol. 159, pp. 651–67.

HINDE, R. A., and SPENCER-BOOTH, Y. (1967), 'The effect of social companions on mother–infant relations in rhesus monkeys', in D. Morris (ed.), *Primate Ethology*, Weidenfeld & Nicolson.

HINDE, R. A., and SPENCER-BOOTH, Y. (1968), 'The study of mother–infant interaction in captive group-living rhesus monkeys', *Proc. Royal Society, B.*, vol. 169, pp. 177–201.

VAN HOOF, J. (1967), 'The facial displays of Catarrhine monkeys and apes'. D. Morris (ed.) *Primate Ethology*, pp. 7–68. Weidenfeld & Nicolson.

VAN HOOF, J. (1971). 'A comparative approach to the phylogeny of laughter and smiling', in R. A. Hinde (ed.), *Non-Verbal Communication*, Royal Society and Cambridge University Press.

HUXLEY, J. S. (1942), *Evolution the Modern Synthesis*, Allen & Unwin.

ISAAC, G. L. (1969), 'Studies of early culture in East Africa', *World Archaeology*, vol. 1, pp. 1–28.

ITANI, J. (1958), 'On the acquisition and propagation of a new food habit in the natural group of the Japanese monkey in Takasaki Yama', *Primates*, vol. 1, pp. 84–98.

ITANI, J. (1959), 'Paternal care in the wild Japanese monkey *Macaca fuscata fuscata*', *Primates*, vol. 2, pp. 61–93.

JAY, P. C. (1963), 'Mother–infant relations in langurs', in H. L. Rheingold (ed.), *Maternal Behavior in Mammals*, Wiley.

JAY, P. (1965), 'The common langur in Northern India', in I. De Vore, (ed.), *Primate Behavior*, Holt, Rinehart & Winston.

JENSEN, G. D., BOBBITT, R. A., and GORDON, G. N. (1968), 'Effects of environment on the relationship between mother and infant pigtail monkeys (*Macaca nemestrina*)', *J. Comparative Physiol. Psychol.*, vol. 66, pp. 259–63.

JOLLY, A. (1966), *Lemur Behavior*, University of Chicago Press.

KAUFMAN, I. C., and ROSENBLUM, L. A. (1967), 'The reaction to separation in infant monkeys: anaclitic depression and conservation withdrawal', *Psychosomatic Medicine*, vol. 29, pp. 648–75.

KAUFMAN, I. C., and ROSENBLUM, L. A. (1969), 'The waning of the mother–infant bond in two species of macaque', in B. M. Foss (ed.), *Determinants of Infant Behaviour IV*, Methuen.

KAUFMAN, J. H. (1965), 'A three-year study of mating behaviour in a free-ranging band of monkeys', *Ecology*, vol. 46, pp. 500–512.

KAWAI, M. (1958), 'On the system of social ranks in a natural troop of Japanese monkeys. I: Basic and dependent rank', *Primates*, vol. 1, pp. 111–30.

KAWAI, M. (1965), 'Newly acquired pre-cultural behaviour of the natural troop of Japanese monkeys on Koshima islet', *Primates*, vol. 6, pp. 1–30.

KAWAMURA, S. (1958), 'Matriarchal social ranks in the Minoo-B troop: a study of the rank system of Japanese monkeys', *Primates*, vol. 2, pp. 181–252.

KLEIN, L. (1971), 'Copulation and seasonal reproduction in two species of spider monkey, *Ateles belzebuth* and *A. geoffroyi*', *Folia primatologica*, vol. 15, pp. 233–48.

KLEIN, L., and KLEIN, D. (1971), 'Aspects of social behaviour in a colony of spider monkeys', *Internat. Zoo Year Book*, vol. 2, pp. 175–81.

KLINGEL, H. (1967), 'Soziale Organization und Verhalten freilebender Steppenzebras', *Zeitschrift für Tierpsychologie*, vol. 24, p. 580.

KOFORD, C. B. (1963), 'Rank of mothers and sons in bands of rhesus macaques', *Science*, vol. 141, pp. 356–7.

KÜHME, W. (1965), 'Freilandbeobachtungen zur Soziologie des Hyänenhundes (*Lycaon pictus*)', *Zeitschrift für Tierpsychologie*, vol. 22, pp. 495–541.

KUMMER, H. (1967), 'Tripartite interactions in hamadryas baboons', in S. A. Altmann (ed.), *Social Communication among Primates*, Chicago University Press.

KUMMER, H. (1968a), *Social Organisation of Hamadryas Baboons*, S. Karger, Basel.

KUMMER, H. (1968b), 'Two variations in the social organization of baboons', in P. Jay (ed.), *Primates*, Holt, Rinehart & Winston.

KUMMER, H., and KURT, F. (1967), 'A comparison of social behaviour in captive and wild hamadryas baboons', in B. H. Vagtborg (ed.), *The Baboon in Medical Research*, University of Texas Press, Austin.

LANCASTER, J. B. (1972), 'Play-mothering: the relations between juveniles and young infants among free-ranging vervet monkeys', in F. Poirier (ed.), *Primate Socialisation*, Random House.

LANCASTER, J. B., and LEE, R. B. (1965), 'The annual reproductive cycle in monkeys and apes', in I. DeVore (ed.), *Primate Behavior: Field Studies of Monkeys and Apes*, Holt, Rinehart & Winston.

LOIZOS, C. (1967), 'Play behavior in higher primates: a review', in D. Morris (ed.), *Primate Ethology*, Weidenfeld & Nicolson.

VAN LAWICK-GOODALL, J. (1968), 'The behaviour of free-living chimpanzees in the Gombe Stream Reserve', *Animal Behaviour Monograph*, no. 1, pp. 161–311.

LINDBERG, D. G. (1972), 'Behaviour and ecology of the rhesus monkey', in A. Roseblum (ed.), *Primate Behaviour, Developments in Field and Laboratory Research, 2*, Academic Press.

LOY, J. (1970a), 'Behavioural responses of free-ranging rhesus monkeys to food shortage', *Amer. J. Physical Anthrop.*, vol. 33, pp. 263–72.

LOY, J. (1970b), 'Peri-menstrual sexual behaviour among rhesus monkeys', *Folia Primatologica*, vol. 13, pp. 286–97.

MARLER, P. (1957), 'Specific distinctiveness in the communication signals of birds', *Behaviour*, vol. 11, pp. 13–39.

MARLER, P. (1968), 'Aggregation and dispersal, two functions in primate communication', in P. Jay (ed.), *Primates*, Holt, Rinehart & Winston.

MARLER, P. (1969), '*Colobus guereza:* territoriality and group composition', *Science*, vol. 163, pp. 93–5.

MASON, W. A. (1968), 'Use of space by *Callicebus* groups', in P. Jay (ed.), *Primates*, pp. 200–216, Holt, Rinehart & Winston.

MAXIM, P. E., and BUETTNER-JANUSCH, J. (1963), 'A field of study of the Kenya baboon', *Amer. J. Physical Anthrop.* vol. 21, pp. 165–80.

MECH, D. (1970), 'The wolf: the ecology and behavior of an endangered species', *New York American Museum of Natural History*, Natural History Press.

MICHAEL, R. P., and HERBERT, J. (1963), 'Menstrual cycle influences on grooming behavior and sexual behavior in the rhesus monkey', *Science*, vol. 140, pp. 500–501.

MICHAEL, R. P., and KEVERNE, E. B. (1970), 'Primate sex pheromones of vaginal origin', *Nature*, vol. 225, pp. 84–5.

MILLER, G. S. (1931), 'The primate basis of human sexual behaviour', *Quart. Review Biol.*, vol. 6, pp. 379–410.

MITCHELL, G. D. (1968), 'Attachment differences in male and female infant monkeys', *Child Development*, vol. 39, pp. 611–20.

MITCHELL, G. D., HARLOW, H. F., GRIFFIN, G. A., and MØLLER, G. W. (1967), 'Repeated maternal separation in the monkey', *Psychoneural Sci.*, vol. 8, pp. 197–8.

MITCHELL, G. D., and STEVENS, C. W. (1969), 'Primiparous and multiparous monkey mothers in a mildly stressful social situation: first 3 months', *Devel. Psychobiol.*, vol. 1, pp. 280–86.

MORRIS, D. (1957), '"Typical intensity" and its relation to the problem of ritualization', *Behaviour*, vol. 11, pp. 1–12.

MOYNIHAN, M. (1964), 'Some behaviour patterns of platyrrhine monkeys. I. The night monkey (*Aotes trivirgatus*)', *Smithsonian Miscellaneous Collection*, vol. 146, p. 5.

MOYNIHAN, M. (1970), 'Some behaviour patterns of platyrrhine monkeys. II. *Saguinus geoffroyi* and some other tamarins', *Smithsonian Contributions to Zool.*, vol. 28, pp. 1–77.

NAPIER, J. R., and WALKER, A. C. (1967), 'Vertical clinging and leaping – a newly recognized category of locomotion behaviour in primates', *Folia Primatologica*, vol. 6, pp. 204–19.

OWEN, D. F. (1969), 'The migration of the yellow wagtail from the Equator', *Ardea*, vol. 57, pp. 77–85.

POIRIER, F. E. (1969), 'The Nilgiri langur (*Presbytis johnii*) troop: its composition, structure, function and change', *Folia Primatologica*, vol. 10, pp. 20–47.

PRAKASH, I. (1962), 'Group organization, sexual behaviour and breeding season of certain Indian monkeys', *Japanese J. Ecology*, vol. 12, p. 83.

RALLS, K. (1971), 'Mammalian scent marking', *Science*, vol. 171, pp. 443–9.

RANSOM, T. W. (1971), 'Ecology and behaviour of the baboon, *Papio anubis*, at the Gombe Stream National Park Tanzania', Ph.D. thesis, psychology department, Berkeley University.

RANSOM, T. W., and RANSOM, B. S. (in press), 'Adult male–infant relations among baboons (*Papio anubis*)', *Folia Primatologica*, vol. 16, pp. 179–95.

RANSOM, T., and ROWELL, T. E. (1972), 'Early social development of feral baboons', in F. Poirier (ed.), *Primate Socialization*, Random House.

RIPLEY, S. (1967), 'Intertroop encounters among Ceylon grey langurs (*Presbytis entellus*)', in S. A. Altmann (ed.), *Social Communication among Primates*, Chicago University Press.

ROSE, R. M., HOLADAY, J. W., and BERNSTEIN, I. S. (1971), 'Plasma testosterone, dominance rank and aggressive behaviour in male rhesus monkeys', *Nature*, vol. 231, pp. 366–71.

ROSENBLUM, L. A., KAUFMAN, I. C., and STYNES, A. J. (1964), 'Individual distance in two species of macaque', *Animal Behaviour*, vol. 12, pp. 338–42.

ROWELL, T. E. (1962), 'Agonistic noises of the rhesus monkey', *Symposium zool. Soc.*, London, vol. 8, pp. 91–6.

ROWELL, T. E. (1963), 'Behaviour and reproductive cycles of female macaques', *J. Reproduction and Fertility*, vol. 6, pp. 193–203.

ROWELL, T. E. (1965), 'Some observations on a hand-reared baboon', in B. M. Foss (ed.), *Determinants of Infant Behaviour*, Methuen.

ROWELL, T. E. (1966a), 'Forest living baboons in Uganda', *J. Zool.*, vol. 147, pp. 344–64.

ROWELL, T. E. (1966b), 'Hierarchy in the organization of a captive baboon group', *Animal Behaviour*, vol. 14, pp. 430–43.

ROWELL, T. E. (1967), 'A quantitative comparison of the behaviour of a wild and a caged baboon group'.

ROWELL, T. E. (1968), 'The effect of temporary separation from their group on the mother–infant relationship of baboons', *Folia Primatologica*, vol. 9, pp. 114–22.

ROWELL, T. E. (1970a), 'Baboon menstrual cycles affected by social environment', *J. Reproduction and Fertility*, vol. 21, pp. 133–41.

ROWELL, T. E. (1970b), 'Reproductive cycles of two *Cercopithecus* monkeys', *J. Reproduction and Fertility*, vol. 22, pp. 321–38.

ROWELL, T. E. (1972), 'Organization of caged groups of *Cercopithecus* monkeys', *Animal Behaviour*, vol. 19, pp. 625–45.

ROWELL, T. E., DIN, N. A., and OMAR, A. (1968), 'The social development of baboons in their first three months', *J. Zool.*, vol. 155, pp. 461–83.

ROWELL, T. E., and HINDE, R. A. (1963), 'Responses of the rhesus monkey to mildly stressful situations', *Animal Behaviour*, vol. 11, pp. 235–43.

ROWELL, T. E., HINDE, R. A., and SPENCER-BOOTH, Y. (1964), 'Aunt–infant interactions in captive rhesus monkeys', *Animal Behaviour*, vol. 12, pp. 219–26.

SAAYMAN, G. S. (1970), 'The menstrual cycle and sexual behaviour in a troop of free-living chacma baboons, *Papio ursinus*', *Folia Primatoligica*, vol. 12, pp. 81–110.

SAAYMAN, G. S. (in press), 'Effects of ovarian hormones upon the sexual skin and mounting behaviour in the free-ranging chacma baboon, *Papio ursinus*', *Folia Primatologica*.

SACKETT, G. P. (1968), 'Abnormal behavior in laboratory reared rhesus monkeys', in M. W. Fox (ed.), *Abnormal Behavior in Animals*, Saunders.

SADE, D. S. (1964), 'Seasonal cycle in size of testes of free-ranging *Macaca mulatta*', *Folia Primatologica*, vol. 2, pp. 171–80.

SADE, D. S. (1965), 'Some aspects of parent–offspring and sibling relationships in a group of rhesus monkeys, with a discussion of grooming', *Amer. J. Physical Anthrop.*, vol. 23, pp. 1–18.

SADE, D. S. (1967), 'Determinants of dominance in a group of free-ranging rhesus monkeys', in S. A. Altmann (ed.), *Social Communication among Primates*, Chicago University Press.

SADE, D. S. (1968), 'Inhibition of son–mother mating among free-ranging rhesus monkeys', *Sci. and Psychoanalysis*, vol. 12, pp. 18–38.

SASSENRATH, E. N. (1970), 'Increased adrenal responsiveness related to social stress in rhesus monkeys', *Hormones and Behaviour*, vol. 1, pp. 283–98.

SEAY, B., HANSEN, E., and HARLOW, H. F. (1962), 'Mother–infant separation in monkeys', *J. Child Psychol. Psychiat.*, vol. 3, pp. 123–32.

SINGH, S. D. (1969), 'Urban monkeys', *Sci. Amer.*, vol. 221, pp. 108–115.

SOUTHWICK, C. H. (1962), 'Patterns of intergroup social behaviour in primates, with special reference to rhesus and howler monkeys', *Annals of the New York Acad. of Sci.*, vol. 102, pp. 436–54.

SPARKES, J. H. (1964), 'Flock structures of the red avadavat with particular reference to clumping and allopreening', *Animal Behaviour*, vol. 12, pp. 125–36.

SPENCER-BOOTH, Y. (1968), 'The behaviour of group companions towards rhesus monkey infants', *Animal Behaviour*, vol. 16, pp. 541–57.

SPENCER-BOOTH, Y., and HINDE, R. A. (1967), 'The effects of separating rhesus monkey infants from their mothers for six days', *J. Child Psychol. Psychiat.*, vol. 7, pp. 179–97.

STOLTZ, L. P., and SAAYMAN, G. S. (1970), 'Ecology and social organization of chacma baboon troops in the Northern Transvaal', *Annals of the Transvaal Museum*, vol. 26, pp. 499–599.

STRUHSAKER, T. (1967a), 'Behaviour of vervet monkeys', *University of California Publications in Zool.*, vol. 82, pp. 1–64.

STRUHSAKER, T. (1967b), 'Social structure among vervet monkeys (*Cerciopithecus aethiops*)', *Behaviour*, vol. 29, pp. 6–121.

STRUHSAKER, T. (1969), 'Correlates of ecology and social organization among African *Cercopithecines*', *Folia Primatologica*, vol. 11, pp. 80–118.

SUGIYAMA, Y. (1960), 'On the division of a natual troop of Japanese monkeys at Takasakiyama', *Primates*, vol. 2, pp. 109–48.

SUGIYAMA, Y. (1964), 'Group composition, population density and some sociological observations of Hanuman langurs (*Presbytis entellus*)', *Primates*, vol. 5, pp. 7–37.

SUGIYAMA, Y. (1965), 'On social change of Hanuman langurs in their natural condition', *Primates*, vol. 6, pp. 381–418.

SUGIYAMA, Y. (1968), 'Social organization of chimpanzees in the Budongo forest, Uganda', *Primates*, vol. 9, pp. 225–58.

TAYLOR, L. (1972), 'Socialization in free-ranging baboons', Ph.D. thesis, Department of Anthropology, Berkeley University.

TSUMORI, A. (1967), 'Newly acquired behavior and social interactions of Japanese monkeys', in S. A. Altmann (ed.), *Social Communication Among Primates*, Chicago University Press.

VANDENBERGH, J. G. (1969), 'Endocrine coordination in monkeys: male sexual responses to females', *Physiol. Behaviour*, vol. 4, pp. 261–4.

VANDENBERGH, J. G., and VESSEY, S. (1968), 'Seasonal breeding of free-ranging rhesus monkeys and related ecological factors', *J. Reproduction and Fertility*, vol. 15, pp. 71–9.

WALKER, A. C. (1967), 'Patterns of extinction among the subfossil Madagascan lemuroids', in *Pleistocene Extinctions, the Search for a Cause*, Proceedings of the 7th Congress of the International Association for Quaternary Research, vol. 6.

WASHBURN, S. L., and DEVORE, I. (1961), 'The social life of baboons', *Scient. Amer.*, vol. 204, pp. 62–71.

WILSON, A. P., and BOELKINS, R. C. (1971), 'Evidence for seasonal variation in aggressive behaviour by *Macaca mulatta*', *Animal Behaviour*.

YOSHIBA, K. (1968), 'Local and intertroop variability in ecology and social behavior of common Indian langurs', in P. Jay (ed.), *Primates*, Holt, Rinehart & Winston.

YOUNG, W. C., and ORBISON, W. D. (1944), 'Changes in selected features of behavior in pairs of oppositely sexed chimpanzees during the sexual cycle and after ovariectomy', *J. Comparative Psychol.*, vol. 37, pp. 107–43.

ZUCKERMAN, S. (1932), *The Social Life of Monkeys and Apes*, Routledge & Kegan Paul.

ZUCKERMAN, S. (1963), 'Concluding remarks of chairman', *Symposium of the zool. Soc.*, London, vol. 10, pp. 119–22.

ZUCKERMAN, S., and PARKES, A. S. (1932), 'The menstrual cycle of the primates, part 5: the cycle of the baboon', *Proc. zool. Soc.*, London, vol. 102, pp. 139–91.

Author Index

Subject Index

Penguin Science of Behaviour

Assessment in Clinical Psychology
C. E. Gathercole

Basic Statistics in Behavioural Research
A. E. Maxwell

Beginnings of Modern Psychology
W. M. O'Neil

Brain Damage and the Mind
Moyra Williams

The Development of Behaviour
W. Mary Woodward

Disorders of Memory and Learning
George A. Talland

Feedback and Human Behaviour
John Annett

The Growth of Sociability
H. R. Schaffer

Interpersonal Perception
Mark Cook

An Introduction to Social Psychiatry
Ransom J. Arthur

Listening and Attention
Neville Moray

Mathematical Models in Psychology
Frank Restle

On the Experience of Time
Robert E. Ornstein

Pathology of Attention
Andrew McGhie

Pattern Recognition
D. W. J. Corcoran

Personal Relationships in Psychological Disorders
Gordon R. Lowe

Psychometric Assessment of the Individual Child
R. Douglass Savage

Schools and Socialization
A. Morrison and D. McIntyre

Socialization
Kurt Danziger

Small Group Psychotherapy
Edited by Henry Walton

The Study of Twins
Peter Mittler

Teachers and Teaching
A. Morrison and D. McIntyre

Vigilance and Attention
Jane F. Mackworth

Vigilance and Habituation
Jane F. Mackworth

Forthcoming

Perception and Cognition: A Cross-Cultural Perspective
Barbara B. Lloyd

Psycholinguistics: Chomsky and Psychology
Judith Greene

The Puzzle of Pain
Ronald Melzack

Pelican books of related interest

Instinct and Intelligence

The Behaviour of Animals and Man

S. A. Barnett

Strictly scientific, yet always readable and often delightful, this Pelican is an introduction to ethology – the growing science of animal behaviour. In it the author of *The Human Species* deploys an intimate knowledge of new research as he explains the behaviour of animals in movement and in their social relations, in courtship and communication; the anatomy and physiology of the nervous systems of animals (from protozoa to primates); the difference between instinct and intelligence and the place of reward and punishment in the training of animals; and, finally, the sources of human behaviour.

'Invaluable to anyone who is interested in the study of animal behaviour, including that of Man himself. The ways in which behaviour patterns may have evolved are discussed, and at all stages comparisons are made between the behaviour of man and that of other animals, although, as the author frequently indicates, great care must be taken to avoid simple analogies. The text is both technical and readable, and is also enlivened by many appropriate quotations from literary sources' *The Times Educational Supplement*

'Serves an extremely useful purpose well . . . an excellent introduction for laymen or for biologists at the first-year undergraduate or sixth-form level' *Science Journal*

The Science of Animal Behaviour

P. L. Broadhurst

For generations men have employed dogs and hawks to hunt, cormorants to fish, and performing animals for entertainment. Modern research, on scientific lines, may greatly widen the use of animals in human society. In this brief and fascinating study the Professor of Psychology at the University of Birmingham recounts how, with the use of test apparatus, monkeys can learn to work for wages paid in token coins; how white rats can be trained to thread their way through a maze or taught specific drills in such devices as the 'shuttle box'. He describes, too, the scientific observations which have been made on the behaviour in the wild of – for instance – penguins or crabs, and the questions that these raise.

Such experimentation and observation, under approved conditions, can be shown to advance the treatment of human mental disorders and to help in the study of such difficult problems as pre-natal influences. The study of animal behaviour may also, as the author suggests, lead to such extraordinary developments as the training of chimpanzees as engine-drivers or the employment of pigeons as production-line inspectors.

This authoritative book explains very clearly the meaning and purpose of modern research into animal behaviour.